JANCIS ROBINSON'S GUIDE TO WINE GRAPES

Oxford   New York

OXFORD UNIVERSITY PRESS

1996

Oxford University Press, Walton Street, Oxford OX2 6DP

Oxford New York

Athens Auckland Bangkok Bogota Bombay
Buenos Aires Calcutta Cape Town Dar es Salaam
Delhi Florence Hong Kong Istanbul Karachi
Kuala Lumpur Madras Madrid Melbourne
Mexico City Nairobi Paris Singapore
Taipei Tokyo Toronto

and associated companies in
Berlin Ibadan

Oxford is a trade mark of Oxford University Press

British Library Cataloguing in Publication Data

Data available

Library of Congress Cataloging in Publication Data

Robinson, Jancis.
[Guide to wine grapes]
Jancis Robinson's guide to wine grapes. p. cm.
1. Grapes—Varieties. 2. Viticulture. 3. Wine and wine making. I. Title.
634.8'3—dc20 SB388.R585 1996 96–14301

ISBN 0–19–866232–7
ISBN 0–19–860098–4 (USA and Australia only)

1 3 5 7 9 10 8 6 4 2

Typeset by Selwood Systems
Printed in Hong Kong

Contents

WHY ALL THESE GRAPES?

The people who make, sell and drink wine are more conscious of the grapes responsible for this uniquely sensual drink than ever before. More and more producers have become aware of the importance of which grape variety grows where and are increasingly trying out new combinations of grapes, grape varieties new to a particular area, or deliberately searching out old vines, capable of producing top quality juice, of grape types that may be obscure or archaic but may be well worth cherishing. They can also choose from the scores of new grape varieties that have been deliberately developed by grape breeders during the last 100 years, designed for specific purposes and conditions.

An increasing proportion of wine bottles are therefore labelled, whether on the front or back, with the names of the grapes whose fermented juice they contain, information of enormous help and increasing interest to us consumers. But for the hundreds of wines that are labelled solely geographically, such as those of which France is mostly proud, a unique alphabetical guide to the grapes behind these names is provided at the back of this guide. With its help, this book should be able to provide at least a few clues as to how any given wine is likely to taste once the cork has been so irrevocably drawn.

There are more than 800 grape names in this book, many of them the aliases under which grapes travel in various different parts of the wine-producing world. No small part of the thrill, and frustration, that is part of assembling a guide like this are the intriguing mysteries of vine identification.

This book is a celebration of the diversity that, despite the ever-present threat of growing internationalization, still exists in the world's vineyards, an attempt to unravel some of its many secrets, and an exhortation to wine lovers everywhere to cherish one of wine's greatest gifts, variety.

GRAPES DO MATTER

Inside every bottle of wine is the fermented juice of hundreds of grapes. Does it matter what sort of grapes these are? Emphatically, yes.

All sorts of things influence how a wine tastes.

♉ The precise weather (and sometimes economic) conditions of the year in which the grapes were picked (the **vintage year**)—and of the year before it.

♉ The way the wine was made (**vinification**) and looked after before it was bottled (**élevage**).

♉ How the vines are planted, grown, pruned, shaped, and treated (**viticulture**).

♉ And, most importantly, the exact environment in which the vines were grown:

  ♉ the **soil** and **subsoil**, particularly its implications for drainage and fertility;

  ♉ the lie, aspect, angle and altitude of the land (**topography**);

  ♉ the general climate of the region (**macroclimate**);

  ♉ the specific local environment of the vineyard which can affect how the grapes ripen, such as nearby lakes or forests (**mesoclimate**);

  ♉ and even the physical conditions of each individual vine (**vine microclimate**).

The French, maddeningly, sum up all of these soil and climate elements in one elegant, mysterious, and untranslatable word: **terroir**.

Many of the finest wines of the world manage to express those local conditions, that terroir, much more eloquently than mere grape variety. They are made from the varieties of grapes that have proved themselves over centuries to be perfectly adapted to that particular environment and their makers see those grapes as a means of communicating the spirit of the place rather than being the message itself. Such wines constitute at most 5 per cent, the most expensive 5 per cent, of all the wine made in the world. They, and their less successful neighbours, tend to rely entirely on **geographical labelling**, selling wines on the basis of the names of the places where they were made. Bordeaux, Soave, and Rioja are three common examples.

It is very difficult, if not impossible, for ordinary wine drinkers to be able to tell very much about any of the factors described above apart from the vintage year. And even if they did, by dint of careful background research, find out exactly which way the vineyard faced and the proportion of oolitic limestone underneath it, it is unlikely this would tell them very much about how the wine will taste.

The same is not true, however, of one more crucial influence on wine's flavour:

♉ The **grape variety(ies)** that go into the wine.

Different vines have different leaves, bunches of different sizes and shapes, and very different-tasting grapes, which produce discernibly different sorts of juice, and therefore wine. This is why an increasing proportion of wine sold today trumpets the name(s) of the grape(s) from which it is made on the front or back label. In fact almost all of the wine made in so-called New World countries is sold on the basis of its 'varietal' make-up, as a Chardonnay, for example, or a Cabernet Sauvignon, a Pinot Noir—or a blend of grape varieties as in Semillon-Chardonnay, Cabernet Sauvignon-Shiraz, or Merlot-Pinot Noir.

Varietal labelling, selling wine on the basis of the grape(s) from which it is made, was introduced in the United States in the mid 20th century and popularized during the California wine boom of the 1970s. It has clearly struck a chord with the millions of new consumers who are bedazzled by the increasingly complicated range of arcane geographical terms and proprietary names they have to choose from today. By concentrating on the grape names they now have a relatively simple way of categorizing the thousands of wines on offer—made all the simpler since so many wines are made from just a handful of the most popular grape varieties. So popular has this concept become that more and more wines from the Old World also now carry some clues as to their varietal make-up, whether it be a Vin de Pays made from a single grape flaunted on the front label, or a geographically named wine whose back label spells out the grape varieties which went into it.

Understanding the likely characteristics of the grape variety from which a wine is made provides a great short cut to making sense of wine. Of course the local environment, or terroir, can play an extremely important part in shaping the character of a wine, but anticipating the likely flavour and characteristics of the grapes—the basic ingredients—in it offers invaluable help in whittling down the possible options when choosing wine. And increased awareness of grape varieties (which was minimal even as recently as the last century) is leading to much more experimentation with unfamiliar vines, even in some of the most hallowed spots on the wine globe.

HOW VINE VARIETIES VARY

Different varieties look different. They are distinguished not just by the different sorts of grapes and wine they produce but by how they look. This has given rise to the science of ampelography, or vine description, whereby the physical characteristics of different varieties are minutely described, not just those of ripe grape berries (which are difficult to preserve and are available for only a week or two each year) but particularly those of leaves, which are easiest to study and can be minutely categorized. Several 19th century scientists wrote monumental ampelographies. Viala and Vormorel's listing of all observed varieties and their synonyms, written between 1902 and 1910, runs to seven large volumes. Dalmasso and Cosmo's Italian version (1960 to 1966) encompasses five. The father of modern ampelography is Pierre Galet, who taught at the University of Montpellier. He has amassed an unparalleled corpus of knowledge of French vine varieties and his disciples Paul Truel and Jean-Michel Boursiquot have travelled the world trying to identify mystery vines.

Another way by which vines can be classified is in their viticultural characteristics. In their annual growth cycles do they bud, flower, or ripen relatively early or late? Are they particularly vigorous? Do they have a tendency to over-produce unless strictly controlled? Are they especially susceptible or resistant to particular pests, diseases, or hazards such as coulure or very cold winters? Are they best suited to one form of pruning?

And then the different characteristics of the bunches or grapes themselves vary, and can have a significant effect on the wines made from them. Some bunches, such as those of Grenache, have particularly tough stems, making that variety especially difficult for machine harvesting. More direct consequences for the sort of wine produced result from particularly small, thick-skinned grapes (lots of colour and flavour) or those with a higher-than-average number of seeds (high tannins).

But to wine drinkers the most important way in which the hundreds of grapes used in wine-making differ is the quite extraordinary, thrilling way in which they each have their own distinct, tasteable personality. The aim of the alphabetical directory which makes up the bulk of this book is to celebrate and elucidate those delicious and intriguing differences.

BLENDING

A wine made exclusively from one grape variety is neither superior nor inferior to one made from a blend of two or more. Some of the finest wines of the world (Château Latour and Dom Pérignon) are always made from a blend whereas others (Château Pétrus and Romanée-Conti) are effectively made from a single grape (Merlot and Pinot Noir respectively).

There are certain classic blending formulae:

Most of the great red bordeaux of the Médoc and Pessac-Léognan (once part of Graves) are made from a blend dominated by Cabernet Sauvignon and supplemented by varying proportions of Merlot, Cabernet Franc, sometimes Petit Verdot, and occasionally some Malbec. This is often known as the Bordeaux blend, the Cabernet/Merlot blend or, in California, as a red Meritage blend. It is worth noting that this recipe evolved not just because the varieties are necessarily complementary in flavour or structure (Merlot nicely adds flesh to Cabernet's framework) but because the varieties are viticulturally complementary; if the flowering, fruit set, or final ripening period of one is unsatisfactory, then others can usually compensate because they flower and ripen at different times.

The white equivalent, that used for the great dry whites of Pessac-Léognan and the sweet whites of Sauternes, is Sémillon and Sauvignon Blanc (together with a little Muscadelle for sweet wines). Californians call this white Meritage and many producers the world over fill in the slight hole in the middle of their Sauvignon Blanc with a little Sémillon ballast.

The traditional champagne recipe is Pinot Meunier, Chardonnay, and Pinot Noir (in that order in the Champagne vineyards currently). Again, expediency plays a part. Pinot Noir provides weight and longevity. Chardonnay brings delicacy, acidity, and aroma. Great wines could be made from these two but Pinot Noir cannot be reliably ripened in the cooler parts of the region, so earlier ripening Meunier, less prone to coulure, is planted instead, and provides wines which are very fruity in youth. This combination is faithfully replicated by producers of top quality sparkling wine outside the Champagne region, although Meunier is often omitted in those parts of California and Australia hot enough to ripen good crops of Pinot Noir.

Then there is the classic Rhône recipe, or rather a traditional licence to blend, wherever they are grown, all manner of the many grapes grown in the Rhône and Languedoc: Syrah, Grenache Noir, Mourvèdre, Cinsaut, and to a certain extent Carignan. A white version of this is also evolving, with interesting blends of any permutation of Viognier, Marsanne, Roussanne, Rolle, and Grenache Blanc emerging not just from the south of France but also from California and Australia.

More and more new blends are appearing in the market-place however, not all of them driven purely by the likely taste of the resulting wine. Growers have to struggle to keep up with the fashion-driven fortunes of individual grape varieties. It takes a good three years until new vines produce anything like a commercial crop (unless field grafting is used). As grapes wax and wane, the market suffers shortages and surpluses of individual varieties, which can spawn some commercially expedient blends. Semillon Chardonnay, or Sem-Chard, is one of these (Semillon being unjustly unfashionable but acceptable as a vehicle for getting the magic C-word on to a label). Colombard Chardonnays, Chenin Blanc Chardonnays, and even Chenin Colombard Semillon Chardonnays have resulted.

When more than one grape is cited on a wine label, they should appear in order of their importance in the blend.

THE VINE FAMILY

The vine, which is essentially a climbing grape tree, belongs to a much larger botanical family, the *Vitaceae* family, which includes hundreds of other plants including the virginia creeper (which has rather grape-like little berries and whose leaves look not unlike those of grapevines). Botanists divide this family into more than a dozen genera. They call the **genus** to which all grapevines belong *Vitis*, which further subdivides into many different **species**, the most important of which are listed below according to their geographical origins. Different members of each species are our old friend vine, or grape, **varieties**.

EUROPEAN AND MIDDLE EASTERN SPECIES
Vitis vinifera

NORTH AMERICAN SPECIES
Vitis aestivalis
Vitis berlandieri
Vitis californica
Vitis candicans
Vitis caribaea
Vitis cinerea
Vitis cordifolia

Vitis girdiana
Vitis labrusca
Vitis linecumi
Vitis longii
Vitis monticola
Vitis riparia
Vitis rufomentosa
Vitis rupestris

NORTH AMERICAN MUSCADINE SPECIES
Vitis munsoniana
Vitis rotundifolia

ASIAN SPECIES
Vitis amurensis
Vitis coignetiae
Vitis thunbergii

Almost all of the wine made today is made from the single species *Vitis vinifera*, the only vine species to originate in Europe, or possibly the Middle East (or even Madagascar). *Vinifera* varieties which are therefore sometimes called European vines. Most of the hundreds of varieties included in this book, and all of the most popular wine grapes, are *vinifera*, although there are a number of varieties, mostly the result of deliberate vine breeding programmes, which have American or even Asian genes in their parentage.

A vine variety bred from more than one variety of the same species is called a **crossing**. Encouraged by their relatively cool climate and a wine law that rewards very ripe grapes, the Germans have been fiendish breeders of new crossings, often designed to ripen particularly early or spectacularly well.

New vine varieties bred from members of more than one species are called **hybrids** or, occasionally, interspecific crossings. **American hybrids** are vines bred, whether deliberately or, more often, by accident in the wild, exclusively from members of different American species. **French hybrids** were bred deliberately at the end of the 19th century and beginning of the 20th century from both *vinifera* varieties and American vines, in an attempt to combat mildew or the greatest scourge of the world's vineyards ever known (see below).

Many American vines, particularly those with *Vitis labrusca* genes, produce grapes with a particularly marked, musky flavour, which reminds some sympathetic tasters of essence of *fraises des bois*. This characteristic is termed 'foxy' and is at its most rampantly odiferous in the popular New York State variety CONCORD. Many different species of vine grow wild in the United States, a phenomenon which made a great impression on many early European settlers.

The Muscadine species, found in the south east of the United States and Mexico, is regarded as botanically distinct from the rest. Many Muscadine varieties have particularly round leaves, and produce small bunches of large berries with relatively slippery, musk-scented juice. The only Muscadine variety likely to impinge on the modern wine drinker's consciousness is the SCUPPERNONG vine, and then only locally.

There is variation, however, even within different vine varieties. Some dark-berried vines such as PINOT NOIR, CARIGNAN, ASPIRAN, GRENACHE, SANGIOVESE, and TERRET have over the centuries shown themselves to be particularly prone to **mutation**. They degenerate so as to produce similar varieties with different attributes, notably the colour of their grapes' skin. Many of these have forms called variously in French Noir (black), Gris (grey), Blanc (white), Rose (pink), Vert (green), Rouge (red), and sometimes more. See, for example, PINOT BLANC, PINOT GRIS.

And then among hundreds of individual varieties, different **clones** of that variety have been identified. Much of the viticultural work of the late 20th century has been concerned with picking out specific plants within a variety for their superior performance in some respect, or respects (high yield being all too common a criterion, although others might be resistance to diseases or environments and, fortunately, wine quality) and reproducing thousands of cuttings from this single plant, so-called **clonal selection**.

PHYLLOXERA: THE GREAT VINEYARD SCOURGE

Hardly surprisingly, the Europeans who first settled in North America tried to make wine out of the native American vines that grew there. They were surprised at how very odd they seemed to taste compared with the wines they were used to and eventually imported European vines. Then they could not understand why these *vinifera* vines failed to thrive.

By the mid 19th century, Europeans were regularly crossing the Atlantic and, thanks to the development of special glass cases, some began to bring back botanical specimens. With the benefit of hindsight we now realize that at least one of these specimens must have borne a tiny louse, which was to have catastrophic effects on the vineyards of Europe. American vines must have developed resistance to **phylloxera**, but European vines had none. Vine-growers first in the southern Rhône and subsequently throughout the rest of France and beyond were mystified when in the 1860s whole tracts of vines seemed to be dying inexplicably. The louse, which devastatingly feasts on vine roots, was soon identified, but it took many years, during which the future of European wine production seemed fatally threatened, before a solution was found.

Eventually it was discovered that if *vinifera* vines are grafted on to the roots of a suitably phylloxera-resistant American vine, then the fruit and therefore wine tastes of the *vinifera* variety in question, but the roots are much more likely to be impervious to the predations of phylloxera.

Parts of the wine-producing world such as South Australia and Chile are yet to be invaded by phylloxera and here vine-growers save themselves considerable time and expense by planting vine cuttings straight into the soil instead of making thousands of grafts and waiting for them to take. In relatively new wine regions such as New Zealand, Oregon, and England, where phylloxera is known but has yet to devastate the local wine industry, many growers decide simply to save money and take the risk of planting ungrafted vines. Phylloxera is known not to thrive in sandy soils, which has saved considerable expense in certain small pockets of vineyard.

ROOTSTOCKS

The great majority of wine, however, is made from *vinifera* vines grafted on to specially bred **rootstocks**, which nowadays can offer an array of different properties. In an area prone to spring frosts, for example, growers may choose a particular rootstock to delay the growth of a variety that buds early, while in very fertile soils a rootstock which naturally limits the amount of leaf growth would be a sensible choice for those growers concerned to maximize quality rather than quantity. (Until the 1980s, quantity was the watchword; nowadays most growers would like to maximize both quantity and quality.) It is also possible to choose rootstocks specifically that offer resistance to nematodes (a type of tiny worm) where these infest the vineyard; or to help resist drought in very dry climates; or to help minimize the problems associated with an excess of lime or salt in the soil. Common rootstocks, usually numbered and named after those who bred them in the late 19th or early 20th centuries, include 1613, 161–49, 1616, and 3309 Couderc; Dog Ridge; 333 École de Montpellier; Fercal; Harmony; 5BB Kober; 41B, 101–14, and 420A Millardet; 1103 Paulsen; 99 and 110 Richter; Riparia Gloire de Montpellier; 140 Ruggeri; Rupestris St George; Schwarzmann; SO4; and 5C Teleki.

All rootstocks have some American vines in their genetic make-up but different American species vary in their resistance to phylloxera. The most resistant species are *Vitis riparia*, *rupestris*, and *berlandieri* so the most resistant rootstocks contain nothing but genes from these species. Some rootstocks with European genes, such as AXR 1 or Aramon Rupestris, which is a *vinifera–rupestris* hybrid and was planted in so many of northern California's newer vineyards because of its high productivity, have proved to have disastrously low resistance.

VINE PESTS AND DISEASES

The phylloxera louse may be the most famous vine pest but it is far from being the only one. Birds, mites, nematodes, and animals as varied as rabbits and kangaroos can harm young vines and both leaves and fruit, damaging the ripening process and the grapes themselves. Numerous insects like nothing more than to prey on juicy vine leaves or fruit. Perhaps most devastatingly, however, insects are responsible for the spread of several serious vine diseases such as leafroll virus and the dreaded Pierce's disease.

The quality and quantity of wine produced by a vine can be seriously affected by a wide range of vine diseases, whether caused by bacteria, a fungus, or a virus. Among diseases spread by bacteria, bacterial blight and crown gall are important, but there is no known cure for Pierce's disease, which kills vines in California and Central and South America in vineyards close to water, where leaf hoppers are endemic. Vine diseases spread by a fungus are by far the most common, although, thanks to the agrochemical industry, they are treatable. Before it succumbed to phylloxera, the vineyards of Bordeaux had been severely attacked by **powdery mildew**, or oidium, another import from North America to which many common European vine varieties are particularly susceptible (including PINOT NOIR, MERLOT, RIESLING). It was found that spraying vines with sulphur keeps powdery mildew in check. **Downy mildew**, or *peronospera*, arrived in Europe from North America soon after phylloxera and continues to pose a summertime threat in wine regions as humid as most of northern Europe. August rains in a French vineyard almost invariably pressage spraying with fine blue copper sulphate, so-called Bordeaux mixture. Among many other serious fungal diseases are **esca** and **Eutypa dieback**, both of which harm vines by rotting their wood.

Perhaps the fungal disease with the most obvious effects for wine drinkers is the **botrytis** fungus. If it affects less-than-ripe grapes in unfavourable weather conditions it is known as **grey rot** and can ruin a vintage, especially red grapes because it spoils the pigments. Grapes, particularly thin-skinned ones, can split and the juice starts to oxidize, losing its freshness and much of its flavour. Thin-skinned grapes in densely packed, compact bunches such as MERLOT, PINOT NOIR, SÉMILLON are some of the most vulnerable.

If, however, the botrytis fungus attacks fully ripe, healthy white grapes in weather conditions which favour its benevolent development into **noble rot** (misty mornings followed by warm, sunny afternoons), then the fungus weaves a magic spell. While apparently transforming bunches into bundles of ash, it concentrates the sugars and develops all sorts of special qualities to make very sweet, wonderfully long-lived wines described as **botrytized** or **botrytis-affected**.

Grafting one plant on to another has effectively doubled the risk of vine diseases transmitted by virus-affected plants. Since grafting became the norm, the effects of virus diseases on the wine-producing world have grown significantly. These effects are mainly economic and usually more obvious to producers than consumers (although many vineyards' beautiful autumn tints owe more to leafroll virus than most wine tourists realize). The other common virus disease which can damage vines and, eventually, the quality of the wine they make is fanleaf degeneration, a particular threat to CABERNET SAUVIGNON.

Different varieties vary in their susceptibility to all these diseases, as they do to those spread by tiny mycoplasma such as flavescence dorée to which CHARDONNAY and RIESLING are particularly susceptible.

Drinking wine is easy. Growing healthy grapes is not.

THE TOP 20 MOST PLANTED WINE GRAPES

It would be impossible to rank grape varieties according to how much wine they produce, as this varies so much from year to year, according to very different influences in different parts of the world. It would also be impossible to discover how many plants of each vine variety are planted since vines are planted to extremely different densities all over the world. Vineyard censuses instead count the area of vineyard planted to each variety. This table, based substantially on vineyard statistics assembled by Patrick W. Fegan of the Chicago Wine School, therefore ranks varieties according to the total amount of the world's surface devoted to each. It is not surprising, then, that Spain's most important varieties play such an important part at the top of this league table; vine densities are exceptionally low in south and central Spain, where summers are notoriously dry and where irrigation has until recently been prohibited. A further complicating factor is that the Spaniards are not the most punctilious in updating their statistics regularly and reliably—a characteristic shared with the Russians, about whose vineyards some educated guesses have been made. The most recent Russian vineyard statistics available are dated 1989.

Grape (synonyms in brackets)	Area in ha/acres (estimated)	Principal countries
Airén W	423,100/1,045,000	Spain
Garnacha R (Grenache, Cannonau)	317,500/784,500	Spain, France
Carignan R (Mazuelo, Carignane)	244,330/603,500	France
Ugni Blanc W (Trebbiano)	203,400/502,400	France, Italy
Merlot R	162,200/400,600	France, Italy
Cabernet Sauvignon R	146,200/361,000	France, Bulgaria, USA
Rkatsiteli W	128,600/317,700	Georgia, Russia, Ukraine, Bulgaria
Monastrell R (Mourvèdre, Mataro)	117,800/291,000	Spain, France
Bobal R	106,200/262,300	Spain
Tempranillo R (Cencibel, Ull de Llebre, Tinta Roriz)	101,600/250,900	Spain, Portugal
Chardonnay W	99,000/244,400	USA, France, Australia, Italy
Sangiovese R (Nielluccio)	98,900/244,300	Italy, Corsica
Cinsaut R (Cinsault)	86,200/212,900	France, S. Africa
Welschriesling W (Laski/Olasz Rizling, Graševina, Riesling Italico)	76,300/188,400	ex-Yugoslavia, Hungary, Romania
Catarratto W	75,400/186,200	Italy (Sicily)
Aligoté W	71,800/177,400	Russia, Ukraine, Moldova, Bulgaria
Muscat of Alexandria W (Moscatel de Málaga, Gordo, Hanepoot)	66,900/165,300	Spain, Australia, S. Africa
Pinot Noir R (Spätburgunder)	62,500/154,400	France, Moldova, Germany, USA
Sauvignon Blanc W	60,700/150,000*	France, Moldova, Ukraine
Chenin Blanc W (Steen)	53,900/133,100	S. Africa, USA, France

*approximate figure based on the unproven assumption that most of Chile's Sauvignon is not Sauvignon Blanc but all of that growing in eastern Europe is.

acids, a group of chemical compounds which give grape juice and wine its tang and ability to refresh. Most common acids in grape juice are tartaric and malic.

American hybrid, a variety bred from American and European vines. See p. 9.

ampelography, the science of identifying grape varieties by detailed description of the appearance of the vine, especially its leaves.

barrel ageing, keeping a wine in cask between fermentation and bottling so that it stabilizes naturally in the presence of small amounts of air and also absorbs some flavour and possibly tannins from the wood, depending on the barrel's age, size, and duration of the barrel ageing.

barrel fermentation, fermentation in small barrels, common for top quality white wine.

Baumé, Australian unit for measuring grape ripeness.

botrytis, a fungal disease which has a malevolent form, grey rot, and a benevolent one which produces some of the finest sweet white wines, noble rot. See p. 12.

Brix, American unit for measuring grape ripeness.

canopy, the above-ground parts of the vine, especially its leaves.

canopy management, viticultural techniques designed to manipulate the canopy to achieve a specific end, usually optimizing the quantity of grapes and quality of wine.

carbonic maceration, a way of making very fruity red wine for early consumption without deliberately crushing the grapes so that fermentation takes place inside the grape.

chaptalization, a wine-making practice common in cooler climates involving adding sugar to the fermentation vat in order to produce a stronger (not sweeter) wine.

clone, an example of a variety replicated from a particular mother vine specially selected for a particular attribute(s). See p. 10.

coulure, a deficient fruit set which may substantially reduce the size of that year's crop. Just after flowering, an excessive proportion of the nascent berries fall off, often because of unsettled cold, wet weather. Some varieties are more prone than others.

crossing, a variety bred from members of the same species.

cultivar, South African term for vine or grape variety.

downy mildew, a fungal disease. See pp. 11–12.

extract, all the solids in a wine; almost everything other than water and alcohol.

fanleaf, a virus disease. See p. 12.

fermentation, the process whereby sweet grape juice is transformed into alcoholic wine, thanks to the action of yeast.

field grafting, grafting a new variety on to an established rootstock in the vineyard. Increasingly common.

flavour compounds, important and intriguing tasteable compounds that may be present in minute concentrations in grapes' pulp and, especially, skins which, together, are responsible for each variety's flavour. Methoxypyrazines make Sauvignon Blanc taste leafy, for example, while monoterpenes make Muscat, Gewürztraminer, and Riesling taste floral.

foxy, the distinctive taste of the grapes and wine of some American vines, especially *Vitis labrusca* and some of its hybrids. Methyl anthranilate is the (often) offending compound.

French hybrid, a vine variety bred from American and European parents.

fruit set, the early summer phenomenon which immediately follows flowering. As soon as the vine flowers, a proportion of them are fertilized, or 'set', to become berries, and eventually grapes. The higher the proportion, the bigger the crop is likely to be.

grafting, broadly, inserting a section of one plant into another so that they unite and grow as one plant. In a viticultural context, usually grafting a European fruiting vine on to a rootstock, often chosen for its resistance to phylloxera.

ha, hectare, or 2.47 acres.

hl, hectolitre, 100 litres, or 26.4 US gallons.

hybrid, a variety bred from members of different species.

leafroll, a virus disease. See p. 12.

lees stirring, a wine-making practice whereby the lees left after fermentation are deliberately agitated to imbue the wine with more flavour; particularly common in barrel-fermented white wines.

malolactic fermentation, a wine-making process which takes place during or after the main alcoholic fermentation and transforms harsh malic acid into softer lactic acid.

millerandage, an abnormal fruit set in which bunches contain berries of very different sizes because of poor fertilization, often because of unfavourable weather.

must weight, a measure of grape ripeness, or sugar concentration in grapes.

mutation, a genetic development causing an observable change, particularly, in some vine varieties, a change of grape berry colour. See p. 10.

noble rot, the benevolent form of botrytis.

oak, the most common sort of wood used for wine barrels, famous for its affinity with the flavours of many grapes.

Oechsle, a German unit for measuring grape ripeness.

phylloxera, fatal vine pest. See p. 10.

powdery mildew, a fungal disease. See p. 11.

pruning, arguably the most important operation of the vineyard year in terms of wine quality. During winter the vine is cut back leaving a specific number of buds responsible for producing the next year's crop. Although many other factors come into play, low-yielding vines in general tend to produce more concentrated wine.

pulp, the fleshy part of the grape containing most of the water, sugars, and acids in grape juice. Apart from red fleshed Teinturiers, the flesh of all grapes is the same dull grey, no matter what the colour of the grape's skin.

rootstock, plant specially selected to form the root system of a fruiting vine of another variety by grafting. See p. 11.

seed, the part of the grape containing tannins. Care is usually taken not to crush them to avoid bitterness.

skin, a very important part of the grape which contains most flavour compounds, pigments, and tannins—all highly desirable, not to say essential, for red wines but a more debatable ingredient in the white wine-making process.

stem/stalk, woody attachment of grape to bunch, high in often harsh tannins. All or most are usually deliberately eliminated by a mechanical destemmer prior to fermentation.

sugar, the carbohydrates accumulated in the grape during the ripening process which are transformed into alcohol by fermentation.

tannins, cheek-drying, astringent compounds similar to stewed tea in effect on the palate which are found in grape seeds, skins, and stems. They can help preserve red wines while they mature in bottle.

training, encouraging a vine to grow into a specific shape, usually to effect some form of canopy management.

vigour, a vine's natural tendency to sprout forth leaves.

vine density, the number of vines planted per unit of area, an important vineyard parameter.

vinifera, a vine species of European origin, as are almost all the well-known wine-producing varieties. See pp. 8–9.

vintage, either the particular year in which the crop was harvested or the process of harvesting itself.

Vitis, the vine genus. See pp. 8–9.

yield, the amount of wine or grapes produced per unit area, usually measured either as tons/acre, tonnes/ha, or, in much of Europe, hl/ha. Many factors such as pressing regime, grape variety, and style of wine affect the conversion of weight of grapes into volume of wine but 1 ton/acre is very approximately equivalent to 17.5 hl/ha.

To get the most from this book, study front and back labels on wine bottles for clues as to which grape varieties have been used. Where no clues are given, try consulting The Grapes Behind the Names appendix.

Cross-references in the text to other entries are denoted by small capitals. Measurements are given in metric and the United States equivalent. Considerable efforts have been made to provide the most recent statistics available. It should be noted however, that the French national authorities publish a vine census only every decade or so, the last one being dated 1988.

Table or drying grapes are omitted from this book unless they are used to make wine.

Key

🍇   Dark-skinned grapes;
                    usually used for making red wine, but may sometimes make pink, or even white wine if the skins are excluded from the fermentation vat.

🍇   Light-skinned grapes;
                    used for making white wines. The skins are too pale (although they may sometimes be quite deep pink) to produce red wine.

Entries for grape varieties which are synonyms for other entries in the book do not usually have their own grape symbol. The reader should thus refer to the main entry to which they are being directed to check if a grape is dark- or light-skinned.

              Indicates the quality range of wines typically produced using a particular grape variety on a scale, from left to right, of poor to excellent.

              e.g.
produces only the most ordinary wine

produces wines which vary in quality from basic to the finest

Ratings have been omitted for particularly obscure or rare vines.

Note: It is worth remembering that very old vines, even of mediocre varieties, can produce concentrated wines. Thus, the respectable quality of the very best examples of, say, Aramon and Carignan in the Languedoc, is not necessarily a direct indication of the inherent quality of these varieties. Conversely, disappointing examples of wines made from even the finest grape varieties exist as a regrettable testament to carelessness, ignorance, or greed, and sometimes all three.

For
W. I. L.

A

ABOURIOU, obscure vine slowly disappearing from the vineyards of South West France. Still allowed into Côtes du Marmandais reds and a number of the region's Vins de Pays, it achieved some degree of international fame in 1976 when French ampelographer Paul Truel identified cuttings sent to Australia from California as 'Early Burgundy' as Abouriou, but most such vines have now been uprooted to make way for something more marketable.

AGHIORGHITIKO, AGIORGITIKO, noble Greek grape also known as St George, or Mavro Nemeas, *the* vine of the Nemea zone. Its full bodied, assertive wine blends well with other varieties (notably with CABERNET SAUVIGNON grown many miles north in Metsovo to make the popular table wine Katoi) and can also produce good quality rosé. Wine produced by Aghiorghitiko is fruity but can lack acidity. Grapes grown on the higher vineyards of Nemea can yield long-lived reds, however, and the grape is Greece's second most planted red, after XYNOMAVRO.

AGLIANICO, southern Italian vine of Greek origin (the name itself is a corruption of the word Ellenico, the Italian word for Hellenic). Its stronghold is the mountainous provinces of Avellino and Benevento in Campania, and the Potenza and Matera provinces in Basilicata. Scattered traces of this early budding vine variety can also be found in Calabria, Apulia, and on the island of Procida near Naples. The vine seems to prefer soils of volcanic origin and achieves its finest results in Taurasi in Campania and Aglianico del Vulture in Basilicata. These two wines share the deep ruby colour, the full aromas, and the powerful, intense flavours which make the variety, at least potentially, one of Italy's finest, even if the potential has so far been realized only in limited quantities. Total Italian plantings had reached about 14,000 ha/34,500 acres by 1991.

AGLIANO, synonym for ALEATICO.

AIDINI, floral-scented variety grown on Santorini and other Greek islands for blending into mainly dry wines.

AIRÉN, the world's most planted vine variety, thanks to the low vine density of vineyards in central Spain where there were still more than a million acres of it in 1996. Accounting for almost a third of all Spain's vineyards, Airén dominates La Mancha and Valdepeñas, thanks to its unusually good drought resistance. It is the major ingredient in the important Spanish brandy business, and has been blended with dark-skinned Cencibel (TEMPRANILLO) grapes to produce light red wines. It is increasingly vinified as a

19

white wine, however, not just in the traditional way to yield heavy wines often marked by oxidation, but also in the modern temperature-controlled way to yield crisp, neutral dry white wines for early consumption with a certain amount of fruit if drunk very young indeed. In several ways therefore, Airén is the Spanish equivalent of France's UGNI BLANC. Experiments with oak have been just as weird. In southern Spain it is known as Lairén.

🍇 **ALARIJEN,** Spanish grape planted in the Extremadura region.

🍇 **ALBALONGA,** German crossing of RIESLING × SILVANER grown to a very limited extent, mainly in Rheinhessen. Rot-prone, it can produce some vigorous Auslese in good years.

🍇 **ALBANA,** Italian vine made famous by the over-promoted Albana di Romagna. Now widely planted in the Emilia-Romagna region, its chief claim to fame is being mentioned in the 13th century by medieval agricultural writer Petrus di Crescentiis. The thick-skinned Albana Gentile di Bertinoro, the most common clone, results in relatively deep-coloured wines. Although Greco and Greco di Ancona are two of its synonyms, it is unrelated to GRECO di Tufo. A total of about 4,500 ha/11,000 acres of Italy were planted with Albana and its various subvarieties in 1990.

🍇 **ALBARELLO,** rare but interesting grape grown to a limited extent near La Coruña in north west Spain.

🍇 **ALBARIÑO,** Spanish name of the distinctive, aromatic, peachy almost VIOGNIER-like high-quality vine grown in Galicia (and as ALVARINHO in the north of Portugal's Vinho Verde region). The grapes' thick skins help them withstand the particularly damp climate, and can result in wines notably high in alcohol, acidity, and flavour. Albariño is one of the few Spanish white grape varieties produced as a varietal and encountered on labels. Most common in Spain in the Rias Baixas zone where it has become so popular that it accounts for about 90 per cent of all plantings and produces some of the most expensive white wine in Spain. Sometimes blended with LOUREIRO, TREIXADURA, CAIÑO. About 2,000 ha/5,000 acres are planted in total in Spain and Portugal.

🍇 **ALBAROLA,** neutral grape quite widely planted in the Cinqueterre zone of Liguria in north west Italy.

🍇 **ALBILLO,** everyday vine grown, sometimes for table grapes, in central Spain. It produces wines that are neutral in flavour, but which can be quite high in glycerol. Occasionally used to soften the rigorous reds of Ribera del Duero.

A light-skinned grape called **Albilla** is widely grown in Peru.

🍇 **ALCAÑÓN,** light, characterful native of Somontano in north east Spain planted on about 100 ha/250 acres.

🍷 **ALEATICO,** bizarre Italian variety with a strong MUSCAT aroma sometimes called Leatico or Agliano today, and referred to by medieval writer Petrus de Crescentiis as Livatica. It may well be a dark-berried mutation of the classic MUSCAT BLANC À PETITS GRAINS and certainly has the potential to produce fine, if somewhat esoteric, fragrant wine such as Avignonesi's version from Tuscan Maremma. It is also grown in Latium and Apulia, but the variety is becoming increasingly rare. Sweet red Aleatico is one of the few wines to be exported from the island of Elba, and the variety is also grown on the French island of Corsica. Also grown in the ex-Soviet Central Asian republics where sweet reds are relatively common, notably Kazakhstan and Uzbekistan.

🍷 **ALFROCHEIRO PRETO,** undistinguished Portuguese variety which can add useful colour to red wine blends grown in Alentejo, Bairrada, Ribatejo, and, particularly, Dão.

🍷 **ALICANTE BOUSCHET,** often known simply as **Alicante** (which itself is sometimes used as a synonym for GRENACHE), a red-fleshed grape used chiefly for blending and grown widely. The most popular of France's TEINTURIERS, thanks to its unique status as a *vinifera* (European) vine, it declined in importance throughout the 1980s but was still the country's 11th most planted dark-skinned vine with 15,800 ha / 39,000 acres, mainly in the Languedoc but also Provence and cognac country in 1988.

Bred between 1865 and 1885 by nurseryman Henri Bouschet from his father's crossing of Petit Bouschet with the popular Grenache, it was an immediate success. (Why Henri Bouschet should have chosen to name this hugely popular variety after a Spanish city is not known.) Thanks to its deep red flesh, the wine it produced was much, much redder than that of the productive and rapidly spreading ARAMON (and much redder than that of its stablemate GRAND NOIR DE LA CALMETTE). In France at least it has tended to be planted alongside the pale but prolific Aramon to add colour in blends. It is also relatively high yielding and on fertile soils can easily produce more than 200 hl/ha (11.4 tons/acre) of wine with 12 per cent alcohol, if little character.

Alicante Bouschet also played a major role in late 19th and early 20th century viticulture as parent of a host of other Teinturiers, the products almost exclusively of crossings with non-*vinifera* varieties.

Outside France it is perhaps most widely cultivated in Spain, where it is also known as Garnacha Tintorera and covered more than 16,000 ha / 39,000 acres in 1990. It is particularly common in Almansa. Alicante is also grown in Portugal, Corsica, Tuscany, Calabria in southern Italy, former Yugoslavia, Israel, and North Africa; and there were still 2,000 acres / 800 ha of it in California in 1992, mainly in the hot Central Valley. Varietal versions are not unknown and tend to major on alcohol, slightly coarse robust fruit and, naturally, colour.

❦ ALICANTE GANZIN, very deep red-fleshed variety which is a parent of all red-fleshed TEINTURIER grapes.

❦ ALIGOTÉ, Burgundy's other white grape. Very much Chardonnay's underdog, but in a fine year, when ripeness can compensate for its characteristic acidity, Aligoté is not short of champions. Its roots almost certainly lie in Burgundy, where it was recorded at the end of the 18th century.

The vine is vigorous and its yield varies enormously according to the vineyard site. If grown on Burgundy's best slopes on the poorest soils in warmer years, Aligoté could produce fine dry whites with more nerve than most Chardonnays, but it would not be nearly as profitable. Aligoté is, typically, an angular, tart wine short on obvious flavour and usually too spindly to subject to oak ageing.

In the Côte d'Or it is being replaced by the two nobler grape varieties, Chardonnay and Pinot Noir, and there were only 500 ha/1,200 acres left in 1988. It is now largely relegated to the highest and lowest vineyards where it produces light, early maturing wines allowed only the Bourgogne Aligoté appellation and drunk either with simple meals by penny pinchers or, traditionally, mixed with blackcurrant liqueur as a kir. Only the village of Bouzeron, where some of the finest examples are produced, has its own appellation for Aligoté, Bourgogne Aligoté-Bouzeron, in which the maximum yield is only 45 hl/ha (2.5 tons/acre) as opposed to the 60 hl/ha allowed for Bourgogne Aligoté. To the north and south of the Côte d'Or, total plantings of Aligoté have remained relatively stable with 500 ha to the south and 200 ha in Chablis country where it makes everyday dry white wine.

Aligoté is extraordinarily popular in eastern Europe. Bulgaria for example had 2,000 ha/500 acres of it in 1993—far more than France—and presumably prized it for its high natural acidity. Aligoté is even more important in Romania, where its 1993 total of 10,500 ha/26,200 acres makes it the country's fourth most common grape, producing mainly varietal wines on its fertile plains. It is also important, planted on tens of thousands of acres, particularly for sparkling wine, in Russia, Ukraine, Moldova, Georgia, Azerbaijan, and Kazakhstan. Aligoté is also grown in Chile and to a very limited extent in California.

ALTESSE, synonym for the ROUSSETTE of Savoie.

ALVA, occasional name for ROUPEIRO in Portugal's Alentejo region.

❦ ALVARELHÃO, grape planted all over northern Portugal, especially in the Douro valley but also in Tras-os-Montes and Dão. As **Alvarello** it was once grown throughout eastern Galicia. See BRANCELLAO.

❦ ALVARINHO, Portuguese answer to the fine white ALBARIÑO of north west Spain, common in top quality Vinho Verde and one of the few Portuguese grape varieties sold in bottled varietal form.

🍷**AMIGNE,** rarity grown in the Valais region of Switzerland producing heady, rich, perfumed wines. See also PETITE ARVINE and HUMAGNE BLANC.

🍷**AMORGHIANO,** rarity found to a limited extent on the Greek island of Rhodes. May well be related to MANDELARIA.

🍷**ANCELLOTTA,** blending ingredient valued for its deep colour and allowed up to 15 per cent in the hugely successful central Italian export Lambrusco Reggiano. According to the Italian agricultural census of 1990, total plantings of Ancellotta were 4,700 ha/11,750 acres, more than of any single LAMBRUSCO subvariety. The odd impressive varietal is made.

ANSONICA, Tuscan name for INZOLIA.

🍷**ARAGNAN,** old south east French vine thought to have originated in Vaucluse and now occasionally found in Palette. Galet suggests it is in fact OEILLADE.

ARAGONEZ, Portuguese name for Spain's TEMPRANILLO used particularly in the Alentejo region.

🍷**ARAMON,** workhorse variety which underpinned the age of mass production in the Languedoc from the mid 19th to the mid 20th centuries. It was displaced as France's most planted variety only in the 1960s by the hardly nobler CARIGNAN.

Aramon's great attribute, apart from its prodigious productivity (up to 400 hl/ha (23 tons/acre) on the fertile plains), was its resistance to oidium, or powdery mildew, the scourge of France's classic wine regions in the mid 19th century. The variety was taken up with great enthusiasm and rapidly spread over terrain previously considered too flat and fertile for viticulture. In the Hérault department the land under vine more than doubled between 1849 and 1869, to 214,000 ha/528,600 acres, while its annual output of wine quadrupled, to 15 million hl/396 million gal.

Such Aramon vines as are still grown in the south of France tend to be a great age, and can, if planted on relatively infertile soils and pruned extremely severely, produce reasonably concentrated, slightly rustic red. But given half a chance the Aramon vine produces enormous quantities of some of the lightest red wine that could be considered red, often with a blue-black tinge and notably low in alcohol, extract, and character. To render the *rouge* sufficiently *rouge* for the French consumer, Aramon had invariably to be bolstered by such red-fleshed grapes as ALICANTE BOUSCHET. Once French wine blenders had access to the deep, alcoholic reds of North Africa, Aramon's fortunes waned (although it was still France's sixth most planted variety in the late 1980s and there were still 32,000 ha/79,000 acres of it in 1990).

Aramon suffers from the twin disadvantages of budding early and ripening late and is therefore limited to hotter wine regions. A little has been known in Algeria and Argentina and its origins are thought by some to be Spanish.

Aramon Gris and **Aramon Blanc**, lighter-berried mutations, can still be found, particularly in the Hérault.

🍇 **ARBOIS,** one of the Loire's less dynamic varieties grown east of Tours and sometimes found in white Touraine, more often in white Valençay, blended with Sauvignon Blanc or Chardonnay. It is declining in importance but was still the third most important variety of any colour in the Loir-et-Cher *département* in 1988. Often called Pineau Menu or Petit Pineau, it is a vigorous vine whose wines are softer than those of the Chenin Blanc that is more common, and more characterful, in the middle Loire valley.

🍇 **ARINTO,** Portuguese variety most commonly encountered in the tangy, occasionally eye-watering whites of Bucelas in which it must constitute at least 75 per cent of the blend. Also grown in many other parts of Portugal, including Ribatejo. Arinto is most notable for its high acidity and can yield wines which gain interest and, sometimes, a citrus quality, with age. **Arinto Miudo** and **Arinto Cachudo** are subvarieties. As an ingredient in Vinho Verde it is known as Paderná.

Arinto do Dão is a different, less distinguished variety.

🍇 **ARNEIS,** dry, scented speciality of Piedmont in north west Italy. Originally from Roero, where it has sometimes softened the dark NEBBIOLO, it has also been called Barolo Bianco, or white Barolo, by its fans. This Piedmontese native was rescued from near extinction in the 1980s, notably by Vietti and Bruno Giacosa. Light almond flavours can please in young wines but tend to fade fast, especially in fully ripe examples which can lack acid. One or two sweet 'passito' versions have been successful.

🍇 **ARRUFIAC** or **ARRUFIAT,** also known as Ruffiac, historic variety enjoying a modest renaissance in Gascony, South West France. The ingredient, along with PETIT COURBU, in Pacherenc du Vic Bilh which distinguishes it from the otherwise similarly constructed Jurançon, it was cleverly rescued from obscurity in the 1980s by André Dubosc of the Plaimont co-operative. Wines produced are not particularly high in alcohol but have an attractive gunflint perfume.

ARVINE. See PETITE ARVINE.

🍇 **ASPIRAN,** traditional Languedoc variety planted on a quarter of all vineyard land in the Hérault *département* just before phylloxera devastated French vineyards in the late 19th century, but rejected for its lack of productivity. It yields light but perfumed red wine and may be an ingredient in Minervois.

⚘ **ASPRINIO,** speciality of the Naples region making light, often slightly sparkling wines.

ASPROKONDOURA of Greece. See BOURBOULENC.

⚘ **ASSARIO BRANCO,** grape encountered in Portugal, especially in the Dão region, which may be Spain's PALOMINO.

⚘ **ASSYRTICO, ASSYRTIKO,** top quality vine increasingly recognized as such in Greece. Wines have substance, the ability to age, and a lime and honeysuckle character. Originally from the island of Santorini, it has an ability to retain acidity in a hot climate which has encouraged successful experimentation with it elsewhere, notably at Domaine Carras of Halkidiki, and it can add nerve to the widely planted SAVATIANO of Attica.

⚘ **ATHIRI,** widely distributed Greek variety whose lemony produce is often used for blending, most notably with the nobler ASSYRTICO.

AUBAINE, very rarely used Burgundian synonym for CHARDONNAY.

⚘ **AUBIN,** almost extinct Mosel variety.

⚘ **AUBUN,** rather ordinary southern Rhône Valley speciality, in decline but still among France's top 20 red wine vines in the early 1990s. In the vineyard it looks extremely similar to COUNOISE, which is officially approved as an ingredient in many appellations of the southern Rhône, eastern Languedoc, and Provence (notably Châteauneuf-du-Pape). Aubun is given as a synonym for Counoise in the regulations for the Coteaux du Languedoc appellation and is allowed into the Cabardès appellation. The wine it produces is a sort of softer, lesser CARIGNAN. It yields well and buds late, offering good resistance to spring frosts.

Aubun (and Counoise) formed part of the vine collection imported into Australia by early settler James Busby and isolated plantings can still be found there.

⚘ **AURORA,** or Seibel 5279, French hybrid once widely planted in the United States. Adaptable and productive, it ripens early but produces little of distinction.

AUVERNAT BLANC, synonym for CHARDONNAY as **Auvernat Gris** is for MEUNIER around the northern French city of Orléans.

AUXERROIS, name used for MALBEC in Cahors where it is the dominant vine variety.

⚘ **AUXERROIS,** also known as **Auxerrois Blanc de Laquenexy,** relatively important variety in north east France (notably Alsace) and Luxembourg, very similar to PINOT BLANC.

Galet, the French ampelographic authority, refutes 19th century suggestions that the variety has some connection with either Chardonnay, Sylvaner, or Melon and maintains it is a distinct variety originally studied at the Laquenexy viticultural station near Metz on the Moselle in the far north east of France. There are still minuscule plantings of Auxerrois in the Loire but today it is most important in Alsace, the French Moselle, and Luxembourg where it is most valued particularly for its low acidity. If yields are suppressed, which they rarely are, the variety can produce excitingly rich wines in both Moselle regions that are worth ageing until they achieve a bouquet with a honeyed note like that of mature Chablis, the wine which today could be described as 'from Auxerre' or, in French, Auxerrois.

Auxerrois is the *éminence grise* of Alsace, where it is much more widely planted than any one of the three true PINOTS planted there. Rarely seen on a label, it produces slightly flabby, broad wines which are blended in to add substance if not subtlety to many a Pinot Blanc. Technically an Alsace wine labelled 'Pinot Blanc' could contain nothing but Auxerrois, which is also a major ingredient in Edelzwicker.

A little Auxerrois is still planted in Baden in Germany and the variety achieved brief international fame in the 1980s when it was discovered that some Chardonnay cuttings smuggled into South Africa in the early 1980s were in fact nothing more glamorous than Auxerrois.

Auxerrois Gris is a synonym for PINOT GRIS in Alsace while Chardonnay, before it became so famous, was once known as **Auxerrois Blanc** in the Moselle.

Gros Auxerrois was a synonym for the dark-berried VALDIGUIÉ.

⚘**AVESSO,** Iberian variety used for Vinho Verde. It produces scented, relatively full bodied wine and is most common in the south of the Vinho Verde region. See also JAÉN.

⚘**AZAL BRANCO,** Portuguese vine which brings acidity to Vinho Verde.

⚘**AZAL TINTO** is the dark-skinned version used for tart red Vinho Verde.

B

⚘**BĂBEASCĂ NEAGRĂ,** Romanian variety whose name, meaning grandmother's grape, compares directly with the much more popular FETEASCA or young girl's grape. There were about 5,000 ha/11,700 acres of this variety planted in the mid 1990s, most of them producing light, fruity reds, considerably less 'serious' than Fetească Neagră.

⚘**BACCHUS,** important German crossing bred from a SILVANER × RIESLING crossing and the lacklustre MÜLLER-THURGAU. In good years can reach heady ripeness with powerful herbaceous, elderflower-like aromas and a bit of fruit, all of which makes it useful for blending with Müller-Thurgau. Unlike the more aristocratic and more popular crossing KERNER, however, the wine produced lacks acidity and is not even useful for blending with high-acid musts in poor years since it too needs to be fully ripe before it can express its own exuberant flavours. Bacchus's great allure for growers is that it can be planted on sites on which Riesling is an unreliable ripener and will ripen as early and as productively as Müller-Thurgau. There were 3,500 ha / 8,600 acres of it in Germany in 1992, about a third more than a decade previously, and more than half of this total was in the Rheinhessen, where its substance is valued as an ingredient in basic QbA blends. The Mosel-Saar-Ruwer and Nahe valleys also grow more than 200 ha, almost exclusively as a sort of alternative to the later-ripening MORIO-MUSKAT for blending purposes, while Franken produces some respectable varietal wines from its increasing area of more than 600 ha.

Bacchus is also grown to a limited extent in England, where it can produce relatively full, aromatic wines.

🍇 **BACO.** Like BOUSCHET, Baco was a nurseryman who saw his name live on in the names of some of the most successful of the vine varieties he bred. Baco's specialities were French hybrids and his most successful was **Baco Blanc**, sometimes called **Baco 22A**, which was hybridized in 1898 and was for much of the 20th century, until the late 1970s, the prime ingredient in armagnac. Baco Blanc was developed to replace FOLLE BLANCHE, which could not be grafted easily after the predations of phylloxera. A crossing of Folle Blanche with the sturdy American hybrid Noah, Baco Blanc is now being pulled up fast, however, as the French seek to purge hybrids from their vineyards. New Zealand grew Baco Blanc in some quantity at one time but here too evidence of a hybrid past is being rapidly eradicated.

🍷 **BACO NOIR**, or **BACO 1**, resulted from crossing Folle Blanche with a variety of *Vitis riparia* in 1894 and was at one time cultivated in such disparate French wine regions as Burgundy, Anjou, and the Landes. It has also been widely planted in Canada and the eastern United States where its smokey wines, often very high in extract, are free of foxy flavours and can age quite well.

🍷 **BAGA**, mainstay of the Bairrada region in Portugal, and probably the country's most planted single vine variety for it is also commonly grown in the Dão and Ribatejo regions, sometimes called Tinta Bairrada. The name means 'berry' in Portuguese, and its berries are notable for their thick skins and the high levels of tannins and acidity in the wines it produces (particularly when, as was traditionally the case, there is no destalking). The finest examples can age well. Vinification techniques are still tart but evolving.

BAIYU, Chinese name for RKATSITELI.

BALSAMINA, Argentine name for their SYRAH.

♆ BANAT RIESLING or **BANAT RIZLING,** Romanian speciality grown just across the border with what was Yugoslavia in Vojvodina. Its produce tends to be somewhat heavy, a sort of LASKI RIZLING without the lift.

♆ BARBAROSSA, Italian variety planted, and occasionally made into varietal wine, in Emilia-Romagna and Corsica. According to Burton Anderson the fast-fading Barbarossa of Liguria is not related. In Provence in south east France it is known as Barberoux and is a permitted, if little planted, grape variety in Côtes de Provence. In both languages the name means 'red beard'.

♆ BARBERA, productive and versatile variety challenged only by SANGIOVESE in its many forms as Italy's most planted dark-berried vine. There were nearly 50,000 ha / 123,500 acres of Barbera planted in Italy in 1990, but it has travelled widely, most notably to the Americas.

Barbera ripens relatively late, as much as two weeks after the other 'lesser' black grape variety of Piedmont DOLCETTO— although still in advance of the stately NEBBIOLO. Its chief characteristic is its high level of natural acidity even when fully ripe, which has helped its popularity in hot climates. The variety has been so long and so well adapted throughout Italy that it has developed various regional strains.

In Piedmont, probably its homeland, Barbera accounts for more than half of all the wine produced most years. Wines from the varied sites on which it is planted are usually relatively low in tannins and high in acid, range from light, tart mouthwashes, through essences of youthful fizziness to powerful, intense wines that need extended cellaring, often the result of oak ageing. New oak, pioneered by Bologna in his Bricco dell'Uccellone, adds spiciness to Barbera's rather neutral aromas and the oak's tannins buffer the grapes' acidity.

Some of the best wines come from the hills immediately to the north and south of Alba and Monforte d'Alba in the Barbera d'Alba zone and the area from Nizza Monferrato north west towards Vinchio, Belveglio, and Rocchetta Tánaro among Barbera d'Asti vineyards.

Barbera dominates much of Lombardy, in particular the vineyards of Oltrepò Pavese, where it makes varietal wines of varying quality and degrees of fizziness, some fine and lively, as well as being blended with the softer local Croatina or BONARDA grape. It is a minor ingredient in Franciacorta and is found, as elsewhere in Italy, in oceans of basic vino da tavola.

Barbera is also much planted, and similarly blended with Bonarda, immediately south east of Piedmont in the Colli Piacentini, the hills above Piacenza, of Emilia-Romagna. Here too it is often blended with Bonarda, particularly in the Val Tidone for the DOC red Gutturnio. It is also planted in the Bologna and Parma hills, the Colli Bolognesi and Colli di Parma, where it may also produce a varietal wine which rarely has the concentration of Piedmont's best. Most of central Italy's Barbera plays a minor role in blends with more locally indigenous varieties, not always adding useful acidity, although that is its real purpose in the deep south. The variety is also grown in Sardinia where some argue that the local PERRICONE, or Pignatello, is also Barbera.

Only just outside Italy, Barbera is grown over the border with what was Yugoslavia in coastal Slovenia. Elsewhere in Europe it is barely known, but Italian immigrants took Barbera with them to both North and, particularly, South America. It is quite widely planted in Argentina, where several thousand ha, mainly in Mendoza and San Juan provinces, can produce juicy, deep-flavoured wines worth ageing. The variety is planted to a much more limited extent in the rest of South America. In California, however, particularly in the hot Central Valley where its naturally high acidity and productivity are treasured, there were more than 11,000 acres/4,500 ha in 1994. Among grape varieties hopefully brought to California from Piedmont, Barbera has consistently outperformed the nobler Nebbiolo. In the North Coast counties it has yielded a few memorable wines (notably made by Louis M. Martini), several of the finest of them blended with PETITE SIRAH. However, most of the Napa and Sonoma plantings have disappeared.

A white berried **Barbera Bianca** is also known.

BARBEROUX, see BARBAROSSA.

BAROLO BIANCO, occasional name for ARNEIS.

🍇 **BAROQUE,** intensely local grape variety of Tursan in South West France, once much more widely grown in the region. The wine produced displays the unusual combination of high alcohol and fine aroma, something akin to ripe pears. Galet claims that the variety was brought back from Spain by pilgrims to Santiago de Compostela.

BEAUNOIS, very rarely used Burgundian synonym for CHARDONNAY.

BELI, eastern European prefix meaning 'white', as in Beli Pinot for PINOT BLANC.

🍇 **BELINA,** historic middle European grape making thin, tart wines. Still important in older vineyards of north east Slovenia and parts of Serbia, and Croatia where it also known as Stajerska Belina. In German it is known variously as Weisser Heunisch and Hunnentraube.

🍇 **BELLONE,** very juicy ancient grape grown near Rome. About 3,000 ha/7,400 acres survived in 1990.

BÉQUIGNOL

🍷 **BÉQUIGNOL,** rare Bordeaux vine occasionally found in nether reaches of South West France such as in the wines of Lavilledieu. The wine is light in body but deep in colour.

BERGERON, local name for ROUSSANNE in the Savoie appellation of Chignin.

🍷 **BIANCAME,** ancient vine (possibly a relative of TREBBIANO according to Anderson), commonly planted along the east coast of northern Italy under many different names, including **Bianchello**. There were about 3,000 ha in 1990.

🍷 **BIANCO D'ALESSANO,** Apulian speciality grown on the heel of Italy.

🍷 **BIANCOLELLA,** Campanian variety often blended with more characterful Italian grapes.

🍷 **BICAL,** Portuguese variety grown mainly in Bairrada, and Dão where it is called Borrado das Moscas, or fly droppings. The wines have good acidity and can be persuaded to display some aroma in a few still varietal versions, although the grapes are typically used in blends for sparkling wines.

BLACK MUSCAT, synonym for MUSCAT HAMBURG.

BLANC, BLANCHE, masculine and feminine French adjectives meaning white and therefore a common suffix for light-berried grape variety names.

🍷 **BLANC DE MORGEX,** alpine speciality of the Valle d'Aosta on the Swiss-Italian border well adapted to high-altitude viticulture.

BLANC FUMÉ, French synonym for SAUVIGNON BLANC, notably in the Pouilly-Fumé appellation in the Loire.

BLANQUETTE, occasionally used as a synonym for a range of white wine varieties in South West France, including BOURBOULENC, CLAIRETTE, MAUZAC, ONDENC. The name has also been used for Clairette in Australia.

BLAU or **BLAUER,** adjective meaning 'blue' in German, often used for darker-berried vine varieties. Blauer Burgunder is PINOT NOIR, for example (while Weisser Burgunder is PINOT BLANC).

🍷 **BLAUBURGER,** Austrian red wine grape variety and, like the much more common ZWEIGELT, a crossing made in the 1920s by Dr Zweigelt at Klosterneuburg, in this case of PORTUGIESER and BLAUFRÄNKISCH. In the late 1980s there were about 500 ha / 1,200 acres of it, the majority planted in Lower Austria, where it produces relatively undistinguished light reds.

BLAUBURGUNDER or even **BLAUER BURGUNDER,** German synonyms for PINOT NOIR, although Spätburgunder is more common.

BLAUER SPÄTBURGUNDER, an Austrian synonym for the PINOT NOIR grape.

🍷 **BLAUFRÄNKISCH,** Austrian name for the lively middle European the Germans call Limberger (and Washington State growers Lemberger). From pre-medieval times, it was common to divide grape varieties into the 'fränkisch' whose origins lay with the Franks and the rest, sometimes called 'wälsch', as in WELSCHRIESLING. This is now one of Austria's most widely planted dark-berried varieties producing wines of real character, if notably high acidity, when carefully grown. Its good colour, tannin, and raciness encourage the most ambitious Austrian producers to lavish new oak on it and treat it like SYRAH. Outsiders, however, can find its build reminiscent of, say, the MONDEUSE of Savoie or one of the denser crus of Beaujolais and for many years it was thought to be the Beaujolais grape GAMAY. Bulgarians still call it Gamé, while Hungarians translate its Austrian name more directly as Kékfrankos.

Its Austrian home is Burgenland, where most of its nearly 3,000 ha/7,500 acres are situated, particularly on the warm shores of the Neusiedlersee. Blaufränkisch is also used to add fruit to blends of Cabernet Sauvignon and Pinot Noir. The variety is sometimes called Nagyburgundi, Frankovka in Slovakia, and Vojvodina is one and the same and here can produce lively, fruity, vigorous wines for early consumption. In Friuli in the far north eastern corner of Italy the variety is called Franconia and can yield wines with zip and fruit.

This vigorous vine buds early and ripens late and can therefore thrive only in a relatively warm climate. Spring frost damage is a risk but yields are usually quite high.

🍷 **BOAL,** name of several varieties grown in Portugal but also, most famously, on the island of Madeira, where its name was Anglicized to Bual. Nowadays total plantings of any form of Boal are extremely small, and are almost exclusively **Boal Cachudo**, clustered around Camara de Lobos on the south coast of Madeira.

🍷 **BOBAL,** important Spanish variety which produces deep-coloured red wines and even grape concentrate in Alicante, Utiel-Requena, and other bulk wine-producing regions in south east Spain. It is widely planted, even though not associated with fine wine. It retains its acidity better than the more popular MONASTRELL, with which it is often grown, and is notably lower in alcohol. The only DO wine in which it is allowed to play a part is Utiel-Requena, although this accounts for a tiny proportion of Spain's total of 106,000 ha/262,000 acres.

Bobal Blanco, also known as Tortosí, is still grown to a limited extent in Valencia.

BOMBINO BIANCO

🍇 **BOMBINO BIANCO,** probably the most important white grape in southern Italy, especially in Apulia but also planted in Emilia-Romagna, Latium, Marches, and the Abruzzi, where it is thought to be the true identity of the variety that is so common that it is, confusingly, called Trebbiano d'Abruzzo, even though it is distinctly less acidic than true TREBBIANO. It is usually encouraged to yield so prolifically that its produce is undistinguished, but Valentini's concentrated, long-living 'Trebbiano d'Abruzzo' shows what can be done.

A sign of how unimportant white grape varieties are in southern Italy, however, is that the Italian vineyard survey of 1990 found only 3,700 ha/9,000 acres of Bombino Bianco, hardly more than a tenth of the area planted with the southern red varieties Montepulciano and Negroamaro, for example.

The vine may have originated in Spain. It ripens late and yields extremely high quantities of relatively neutral wine, much of which has been shipped north, particularly to the energetic blenders of Germany. Some of its synonyms, Pagedebit and Straccia Cambiale in particular, allude to its profitability to the vine-grower.

There is also a much less common dark-berried **Bombino Nero** in Apulia.

🍇 **BONARDA,** name of at least three Italian varieties. In Oltrepò Pavese and Colli Piacentini it is the local name for CROATINA. In the Novara and Vercilli hills it is the local name for the UVA RARA used to soften SPANNA. In Piedmont **Bonarda Piemontese** is an almost extinct aromatic, pre-phylloxera variety which some have tried to revive for blending with Barbera. A vine called Bonarda is known in Brazil and is so widely grown in Argentina that South America has six times as much Bonarda planted as Italy. The Argentine authority Alcalde believes his country's Bonarda may well be the same as California's CHARBONO.

🍇 **BONDOLA,** traditional *vinifera* grape of the Ticino, where it may be an ingredient in a local blend called Nostrano (as opposed to Americano, from hybrids). MERLOT is now much more popular.

🍇 **BORBA,** productive Spanish vine grown in Extremadura.

BORDELAIS, South West French name for any variety or varieties thought (often wrongly) to originate in Bordeaux. Could be BAROQUE, COURBU, FER, TANNAT, and a host of others.

🍇 **BORDO,** significant vine for the Brazilian wine industry. Ives Noir is an alternative name. Also, an occasional north east Italian name for CABERNET FRANC.

BORRACAL, synonym for Galicia's CAIÑO TINTO in Portugal's Vinho Verde region.

BORRADO DAS MOSCAS, the Dão region's name for the Portuguese variety BICAL.

🍷 **BOSCO,** very ordinary Ligurian that forms the basis of Cinqueterre and oxidizes easily.

🍷 **BOUCHALÈS,** very minor variety of Bordeaux and the Lot-et-Garonne *département*. Total French plantings fell from over 4,000 ha/9,800 acres in 1968 to less than 500 ha 20 years later.

BOUCHET, the name for CABERNET FRANC used in Saint-Émilion, Pomerol, Fronsac, etc. in Bordeaux.

BOUCHY, Madiran synonym for CABERNET FRANC.

🍷 **BOURBOULENC,** ancient variety that may well have originated in Greece, as the now rarely seen Asprokondoura, and has been grown throughout southern France for centuries. Ripening late but keeping its acidity well, it is allowed into a wide variety of Provençal and southern Rhône appellations (including Châteauneuf-du-Pape) but is rarely encountered as a dominant variety other than in the distinctively marine-scented whites of La Clape in the Languedoc. France's total area planted with Bourboulenc halved in the 1970s and then doubled again, to about 800 ha/2,000 acres, in the 1980s thanks in part to a re-evaluation in the Languedoc, where it is also, confusingly, known as Malvoisie. It is difficult to ripen fully in the southern Rhône, and its tight bunches of large grapes can make it prone to rot in more difficult years but Bourboulenc, together with MACCABÉO and GRENACHE BLANC, is important in any white Minervois or Corbières Blanc.

BOUSCHET, not a grape variety but, like Müller, Scheu and Seibel, a vine-breeder's surname that lives on in the name of his creations, although in this case there were two Bouschets, a 19th century father and son whose work, perhaps unfortunately, made the spread of ARAMON possible. In 1824 Louis Bouschet de Bernard combined the productivity of Aramon with the colour expected of a red wine by crossing Aramon with the red-fleshed Teinturier du Cher and modestly calling the result Petit Bouschet. This expedient crossing was popular in France throughout the second half of the 19th century and is still to be found in parts of North Africa and, according to Galet, Portugal. Louis's son Henri carried on where his father left off, producing most durably ALICANTE BOUSCHET and GRAND NOIR DE LA CALMETTE as well as a Carignan Bouschet.

🍷 **BOUVIER,** minor variety bred as a table grape and now grown mainly in Austria's Burgenland region, where it is particularly used for Sturm, the cloudy, part-fermented sweet grape juice that is a local speciality at harvest time. Also grown in the Mátra Foothills of Hungary.

BOVALE

❦ **BOVALE,** Sardinian grape varieties distinguished as the rather more austere **Bovale Sardo** and the more common **Bovale Grande**. Both are used mainly for blending and may be related to Spain's BOBAL.

❦ **BRACHETTO,** light red-skinned Piedmontese grape found particularly round Asti, Roero, and Alessandria, where it is particularly successful. It produces controversial wines, notably Brachetto d'Acqui, that are fizzy, relatively alcoholic, and have both the colour and flavour of strawberries. Brachetto is the Italian name for **Braquet**, which the French consider an old Provençal speciality and which is a valued ingredient in the red and pink wines of Bellet near Nice. Yields are low and the vine is quite delicate but the wine is truly distinctive.

❦ **BRANCELLAO,** Galician vine once widely grown in north west Spain, particularly in Rias Baixas and known in Portugal's Vinho Verde region as **Brancelho** for its pale but sometimes aromatic wine. It is the ALVARELHÃO of port country, and was known as Alvarello in eastern Galicia.

BRAUCOL, BROCOL, local synonyms for FER in Gaillac.

BRETON, name used in the middle Loire for the CABERNET FRANC grape. The reference is not to Brittany but to Abbot Breton, who is reputed to have disseminated the vine in the 17th century.

❦ **BROWN MUSCAT,** dark-skinned strain of MUSCAT BLANC À PETITS GRAINS used particularly in north east Victoria for the Australian speciality Liqueur Muscat.

❦ **BRUNELLO,** qualitatively important, powerful clone of SANGIOVESE isolated in the mid 19th century in Montalcino in Tuscany, by Ferruccio Biondi Santi. The vigorous vine ripens late, has good disease resistance and smallish bunches of compact, thick-skinned grapes with a brown-tinted skin. Wines are much more flamboyant than the average Sangiovese but the Italian vineyard census of 1990 found only 1,100 ha / 2,700 acres of Brunello (as opposed to 86,200 ha / 213,000 acres of Sangiovese).

❦ **BRUN FOURCA,** ancient Provençal vine still grown to a very limited extent in Palette.

BUAL, Anglicized form of BOAL, name of several Portuguese white grape varieties much planted in the 19th century.

❦ **BUDAI ZÖLD,** Transylvanian speciality grown in Hungary around Lake Balaton, making deep-coloured, full bodied wine mainly consumed locally.

❦ **BUKETTRAUBE,** South African speciality which ripens early to produce unremarkable, vaguely grapey, off-dry wines but can occasionally be persuaded to produce some exceptional botrytized sweet wine.

🍷 **BURGER,** once California's most planted *vinifera* variety, having been promoted by one pioneer as greatly superior to the MISSION grape. Galet identified it as the almost extinct southern French variety Monbadon that was cultivated to a limited extent in the Languedoc until the 1980s. It produces sizeable quantities of neutral wine. In the early 1990s there were more than 2,000 acres/800 ha of Burger, mainly in the hot Central Valley.

BURGUNDAC CRNI, Croatian name for PINOT NOIR while **Burgundac** with or without the suffix **Beli** is usually PINOT BLANC.

BURGUNDER, common suffix in German, meaning literally 'of Burgundy', for such members of the PINOT family as Spätburgunder, Blauer Spätburgunder, Blauburgunder, or Blauer Burgunder (PINOT NOIR); Weissburgunder or Weisser Burgunder (PINOT BLANC); and Grauburgunder (drier styles of PINOT GRIS).

C

CABERNET is loosely used as an abbreviation for either or both of the black grape varieties CABERNET FRANC and (more usually) CABERNET SAUVIGNON. North east Italian wines labelled Cabernet have tended to be mainly Cabernet Franc.

🍷 **CABERNET FRANC,** French variety much blended with and overshadowed by the more widely planted CABERNET SAUVIGNON. Only in Anjou-Touraine in the Loire Valley and on the right bank of the Gironde in Bordeaux (in St-Émilion etc.) is it more important than Cabernet Sauvignon. It will ripen in a much cooler environment than Cabernet Sauvignon and its characteristic scent is akin to pencil shavings.

As a wine, Cabernet Franc tends to be rather lighter in colour and tannins, and therefore earlier maturing, than Cabernet Sauvignon, although the world's grandest Cabernet Franc-dominated wine Ch Cheval Blanc proves that majestic durability is also possible. Cabernet Franc is, typically, light to medium bodied with more immediate fruit than Cabernet Sauvignon and some of the herbaceous aromas evident in unripe Cabernet Sauvignon.

In the vineyard it looks very like the other Cabernet except that it has less dramatically indented leaves but the two share so many characteristics that it seems likely that Cabernet Franc is a particularly well-established mutation of the other Cabernet, well adapted to the right bank's cooler, damper conditions. By the end of the 18th century it was well established as producing high-quality wine in the Libournais vineyards of St-Émilion, Pomerol, and Fronsac where it is often called Bouchet today. In the Loire, as Breton, it takes its name from an abbot who imported it there from the south.

CABERNET FRANC

Cabernet Franc is particularly well suited to cool, inland climates such as the middle Loire and the Libournais. It buds and matures more than a week earlier than Cabernet Sauvignon, which makes it more susceptible to coulure, but it is easier to ripen fully and is much less susceptible to poor weather during harvest. In the Médoc and Graves districts of Bordeaux, where Cabernet Franc constitutes about 15 per cent of a typical vineyard and is always blended with other varieties, it is widely regarded as a form of insurance against the weather's predations on Cabernet Sauvignon and Merlot grapes rather than a positive flavour addition. Most Libournais bet on Cabernet Franc in preference to the later, and therefore riskier, Cabernet Sauvignon to provide a framework for Merlot. The first choice for those replanting white wine vineyards, by 1990 Cabernet Sauvignon was almost twice as popular in the Bordeaux region as Cabernet Franc, planted on a total area of 13,400 ha/33,000 acres.

Cabernet Franc is still planted all over South West France, although Cabernet Sauvignon is gaining ground. If Cabernet Franc was France's eighth most planted black grape variety in 1988, this was largely thanks to its ascendancy in the Loire (where it was the single most planted grape variety with 11,000 ha/27,000 acres, in 1988). Steadily increasing appreciation of relatively light, early maturing reds such as Saumur-Champigny, Bourgueil, Chinon, and Anjou-Villages fuelled demand for Cabernet Franc in the Loire at the expense of the competing Rosé d'Anjou and Chenin Blanc whites.

Cabernet Franc is also well established in Italy, particularly in the north east, where high yields too often result in over-herbaceous aromas and a lack of fruit. The 1990 Italian vineyard survey found nearly 6,000 ha of Cabernet Franc in Italy (as opposed to only 2,400 of Cabernet Sauvignon). It is occasionally called Cabernet Frank or even Bordo, but is more usually labelled simply Cabernet once in the bottle. Italian vine-growers have tended to be as insouciant about the distinction between the two Cabernets as their counterparts over the border in Slovenia and further east. In the early 1990s Cabernet Franc was relatively common in Kosovo and was the only western grape to be grown for red wine production in Albania. Among the ex-Soviet republics only Kazakhstan seems to grow it.

Vine-growers in north west Spain are convinced their MENCÍA is none other than Cabernet Franc.

Elsewhere, Cabernet Franc has until recently been planted merely in order to follow the Bordeaux recipe slavishly (even in very different climates such as that of Napa Valley) but a few varietal versions are appearing. Although there was some early confusion between Cabernet Franc and Merlot, Californians have been rearing Cabernet Franc since the late 1960s, and with zeal since the 1980s. The state's total acreage was 1,700 in 1992, most in Napa and Sonoma counties and one-third of it too young to bear fruit. It is expected to play an increasingly important role here.

In 1990 a third of Australia's 300 ha / 700 acres of Cabernet Franc (like two-thirds of its 400 ha of Merlot) was too young to bear a crop.

The vine is increasingly planted in South America, notably Argentina, where there were 500 ha / 1,200 acres in the early 1990s, most of it in Mendoza.

New World wine regions that have shown a particular aptitude for well-balanced, fruity wines based predominantly on Cabernet Franc include Long Island in New York State and Washington State. New Zealand also shows promise.

CABERNET FRANK, occasional north-east Italian name for CABERNET FRANC.

🌿 CABERNET SAUVIGNON, the world's most renowned grape variety for the production of fine, long-lived red wine. Originating in Bordeaux, particularly the well-drained Médoc and Graves, where this late ripening vine is almost invariably blended with other grapes, it has been taken up in other French wine regions and in much of the Old and New Worlds, where it has been blended with traditional native varieties, with its traditional Bordeaux blending partner MERLOT, but has more often been used to produce pure varietal wine.

Perhaps the most extraordinary aspect of Cabernet Sauvignon is its ability to travel, to set down its roots in distant lands and still produce something that is recognizably Cabernet, whatever the circumstances. And what makes Cabernet Sauvignon remarkable to taste is not primarily its exact fruit flavour—although that is often likened to blackcurrants, its aroma sometimes to green bell peppers—but its structure and its ability to provide the perfect vehicle for individual vintage characteristics and wine-making techniques and local physical attributes, or *terroir*. In this respect it resembles the equally popular and almost as ubiquitous CHARDONNAY, to whose 'vanilla' Cabernet Sauvignon is often compared as 'chocolate'.

It is Cabernet Sauvignon's remarkable concentration of tannins, pigments, and flavour compounds that really sets it aside from most other widely grown vine varieties. It is therefore easily capable of producing deeply coloured wines worthy of long maceration and wood ageing with a demonstrable affinity for densely textured French oak. The particular appeal of Cabernet Sauvignon lies much less in primary fruit aromas than in the much more subtle flavour compounds that evolve over years into a subtle bouquet.

Cabernet Sauvignon's origins remain mysterious, although there have been many wild and some wonderful theories. What is certain is that it was not until the end of the 18th century, when the great estates of the Médoc and Graves were established, that Cabernet Sauvignon started to make any significant impact on the vineyards of Bordeaux.

The distinguishing marks of the Cabernet Sauvignon berry are its small size, its high ratio of pip to pulp, and the thickness of its skins, so distinctively blue, as opposed to red or even purple, on the vine. The pips are a major factor in Cabernet Sauvignon's high tannin level, while the skins account for the depth of colour that is the tell-tale sign of a Cabernet Sauvignon in so many blind tastings. The thickness of the skins also makes the vine relatively resistant to rot.

The vine is susceptible, however, to powdery mildew, which can be treated quite easily, and the wood diseases eutypa and excoriose which cannot. It is extremely vigorous and should ideally be grafted on to a weak rootstock to keep its leaf growth in check. It both buds and ripens late, one to two weeks after Merlot and Cabernet Franc, the two varieties with which it is typically blended in Bordeaux. Cabernet Sauvignon ripens slowly, which has the advantage that picking dates are less crucial than with other varieties (such as Syrah, for example); but this has the disadvantage that Cabernet Sauvignon simply cannot be relied upon to ripen in the coolest wine regions, especially when its energy can so easily be diverted into producing dangerously shady leaves such as in Tasmania or New Zealand unless canopy management techniques are employed. Cabernet Sauvignon that fails to reach full ripeness can taste eerily like Cabernet Franc.

Even in the temperate climate of Bordeaux, the flowering of the vine can be dogged by cold weather and the ripening by rain, so that Bordeaux's vine-growers have traditionally hedged their bets by planting a mix of early and late local varieties, typically in the Médoc and Graves districts 70 per cent of Cabernet Sauvignon plus a mixture of Merlot, Cabernet Franc, and sometimes a little Petit Verdot. (See CABERNET FRANC for reasons why so little Cabernet Sauvignon is planted in St-Émilion and Pomerol.)

A practice that had its origins in canny fruit farming has proved itself in the blending vat. In cooler climates the plump, fruity, earlier maturing Merlot is a natural blending partner for the more rigorous Cabernet Sauvignon, while Petit Verdot can add extra spice (if only in the sunniest years) and Cabernet Franc can perfume the blend to a certain extent. Wines made solely from Cabernet Sauvignon can lack charm and stuffing; the framework is sensational but tannin and colour alone make poor nourishment. As demonstrated by the increasing popularity of Merlot and Cabernet Franc and even Petit Verdot cuttings, newer wine regions have begun to follow the Bordeaux example of blending, although the Médoc recipe is by no means the only one. In Tuscany Cabernet is commonly blended with Sangiovese. In Australia and, increasingly but with very different results, in Provence it is blended with SYRAH (SHIRAZ).

Cabernet Sauvignon is by quite a margin the most planted top quality vine variety in the world (if one excludes Grenache which, in its most common form Garnacha, rarely performs at the peak of its potential). Only those regions such as England, Germany, and Luxembourg, disbarred for reasons of climate, have resisted joining this particular club on any significant scale. Cabernet production has become almost a rite of passage for the modern wine-maker wishing to make his mark.

French plantings of Cabernet Sauvignon increased enormously in the 1980s so that by 1988 there were 36,500 ha/90,000 acres, of which two-thirds were in the Bordeaux region (although the easier-to-ripen Merlot is more popular here). Cabernet Sauvignon, although little planted north of the river Dordogne, is now much more common in the Entre-Deux-Mers between the Gironde and Dordogne rivers as less profitable white varieties have been uprooted. Such is the size of this region that there is more Cabernet Sauvignon planted there, 11,000 ha recorded in 1988, than in any other Bordeaux district, including its Médoc stronghold.

The vine is planted throughout South West France, notably in Bergerac and Buzet. In more internationally styled wines, however, it may add structure to the Malbec of Cahors, the Négrette of Côtes du Frontonnais, the cocktail that is red Gaillac, and the Tannat of Madiran, Côtes de St-Mont, Irouléguy, and Béarn.

Plantings in the Languedoc increased substantially in the 1980s to a total of 11,700 ha/28,900 acres in 1990 but Cabernet Sauvignon has not been nearly so successful here as Syrah, which covered a total of 24,000 ha by the same year. Cabernet Sauvignon does not tolerate very dry conditions without more substantial irrigation than is condoned by the French authorities. The most obviously successful southern French Cabernet Sauvignons are those used as ingredients in low-yield blends with Syrah and other Rhône varieties such as some of the best of Provence.

Cabernet Sauvignon's only other French territory is the Loire, but, despite the freedom allowed by most appellation regulations to choose either Cabernet Franc or Cabernet Sauvignon or both for local reds, most vine-growers prefer the regularity of the former to the risks involved with growing the latter in relatively cool conditions.

The variety is very widely planted in eastern Europe, with some of the most impressive bottled-aged examples coming from Moldova. It is widely planted in Russia and Ukraine, although in Russia's cooler wine regions the cold-hardy hybrid CABERNET SEVERNY is becoming increasingly popular. Cabernet Sauvignon is also grown in Georgia, Azerbaijan, Kazakhstan, Tajikistan, and Kyrghyzstan. The vine is very important in Bulgaria, Romania, and what was Yugoslavia. Even when expected to produce relatively high yields, eastern European Cabernet Sauvignon is unmistakably Cabernet, and the best Romanian and Bulgarian wines have real depth of flavour as well as colour. It is grown to a much more limited extent in Hungary, Austria, and Greece where it was first planted, in modern times at least, at Domaine Carras.

Another country with an important area planted with the world's noblest black grape variety is Chile, whose grand total of (ungrafted) Cabernet Sauvignon is well over 12,000 ha/30,000 acres, making it the country's most important dark-skinned vine variety by far—apart from the rustic PAIS. Here the fruit is exceptionally healthy and the wine, if made carefully in one of the more modern wineries, almost rudely exuberant.

Not surprisingly, Cabernet Sauvignon also flourishes in the rest of South America's vineyards: in Argentina, where in terms of quantity it is dwarfed by MALBEC; in Brazil, Uruguay, Mexico, Peru, and Bolivia.

Cabernet Sauvignon, even less surprisingly, has been the bedrock of that construct called California collectable. In 1994 the state could boast more than 36,000 acres/14,600 ha of it, very slightly less than of the most widely planted black grape variety Zinfandel, but much less than plantings of any of the white wine varieties Chardonnay, Colombard, or Chenin Blanc. In the best North Coast sites, notably as 100 per cent varietals or blended to the Bordeaux recipe as Meritage wines around Rutherford, St Helena, and Oakville in the Napa Valley, it can yield wines of both structure and stature to rival the best of the Médoc.

Cabernet Sauvignon is also one of Washington State's two major black grape varieties, along with the hugely successful Merlot, but the climate is too cool to ripen the vine reliably in many other states. So glamorous is Cabernet's image, however, that even the wine industry in Canada with its natural climatic disadvantages persists with it, to increasingly impressive effect.

If Californians decided early on that the Napa Valley was their Cabernet Sauvignon hotspot, Australians did the same about Coonawarra. They, however, have for decades employed a much less reverential policy towards blending their Cabernet. Cabernet-Shiraz blends (a recipe recommended in Provence as long ago as 1865 by Dr Guyot) have been popular items in the Australian market-place since the 1960s. The richness and softness of Australian Shiraz is such that it fills in the gaps left by Cabernet Sauvignon even more effectively than French Syrah can do. The classic Bordeaux blend is still very much rarer on the other hand, as one might expect from wine producers more determined than the Californians to go their own way independently of Europe.

Cabernet Sauvignon has long played a part in the New Zealand wine industry, and with 475 ha/1,170 acres it was the country's fourth most planted vine variety in 1990. It was not until the mid to late 1980s, however, that Cabernet's tendency to transform into excess foliage rather than ripe fruit was mastered, and fully ripe colours and flavours in New Zealand Cabernets became an attractive reality. Cabernet Sauvignon is also seen as *the* noble red grape variety in South Africa, where there were 2,400 ha in 1990. In North Africa there are very limited plantings in Morocco.

Cabernet Sauvignon has been an increasingly popular choice for internationally minded wine producers in both Portugal and, particularly, Spain, however. In Spain it was planted by the Marqué de Riscal at his Rioja estate in the late 19th century, and could also be found in the vineyards of Vega Sicilia, but was otherwise virtually unknown on the Iberian peninsula until the 1960s when it was imported into Penedès by both Miguel Torres and Jean León. It is slowly broadening its base in Spain, especially in Navarre, not just for wines dominated by it but for blending. And in Portugal it could already be found, blended with indigenous grape varieties, in a handful of lush red wines made in the Lisbon area by the mid 1980s.

Italy, where Cabernet Sauvignon was introduced, via Piedmont, in the early 19th century, now has a very substantial area of Cabernet vineyard, although Italians have been somewhat cavalier about distinguishing between the two very different sorts of Cabernet either on the label or, sometimes, in the vineyard. The 1990 Italian vineyard survey identified just 2,400 ha of Cabernet Sauvignon as compared with a total of nearly 6,000 ha of Cabernet Franc. The variety has played a considerable role in the emergence of Supertuscan blends, and can be found as a seasoning in an increasing proportion of Chianti. It is officially sanctioned, and individually specified, in such DOCs as Carmignano in Tuscany; Colli Bolognesi in Emilia-Romagna; in Trentino; in Lison-Pramaggiore in the Veneto; and in Friuli in Colli Orientali, Collio, Grave del Friuli, Isonzo, and Latisana. The vine is also even making incursions into Piedmont, NEBBIOLO territory.

Perhaps the most tenacious Cabernet Sauvignon grower of all has been Serge Hochar of Ch Musar in the Lebanon, and there are other, rather less war-torn pockets of Cabernet Sauvignon vines all over the eastern Mediterranean in Turkey, Israel, and Cyprus. In the Far East, there have been experiments with the vine in both China and Japan, where its strong links with the famous châteaux of Bordeaux are particularly prized.

Wherever there is any vine-grower with any grounding in the wines of the world, and wherever late ripening grapes are economically viable, Cabernet Sauvignon is almost certain to be grown—except in Bordeaux's great rival regions Burgundy and the Rhône. It is perhaps significant that, in contrast to many other black grape varieties, no white, pink, or grey version of this definitive red wine inspiration exists.

℥ **CABERNET SEVERNY,** cold weather speciality bred at the All-Russia Potapenko Institute in Rostov. It was created by pollination of a hybrid of Galan × *Vitis amurensis* with a pollen mixture of other hybrid forms involving both the European vine species *Vitis vinifera* and the famously cold-hardy Mongolian vine species *Vitis amurensis*.

CADARCĂ, Romanian name for Hungary's KADARKA.

CAGNINA, possible synonym for REFOSCO in Italy's Romagna region.

℥ **CAIÑO TINTO,** strongly perfumed, delicate Galician variety found in the tart red wines of Rias Baixas and Ribeiro. Known as Borraçal in Portugal's Vinho Verde country.

CALABRESE

The light-berried **Caiño Blanco,** is found particularly in Rias Baixas.

CALABRESE, meaning 'of Calabria', is a common synonym for NERO D'AVOLA.

🍇 **CALADOC,** GRENACHE-like crossing created by French ampelographer Paul Truel from Grenache and MALBEC to produce a vine less prone to coulure. It has been planted to a limited extent in the southern Rhône but is not allowed into any appellation contrôlée wine, although it may be used to add tannin and aroma to red vins de pays in Provence.

🍇 **CALAGRAÑO,** historic and nearly extinct Rioja grape which can make astringent but interesting wine, according to Radford.

🍇 **CALITOR,** almost extinct, naturally productive ancient Provençal vine allowed but rarely used in wines of the southern Rhône. Its wines tend to be light, but in hillside sites they can add character. May be the Catalan Garriga.

🍇 **CALLET,** rustic Majorcan grape often planted mixed in the vineyard with Fogoneu Mallorquí or Fogoneu Francés. All three tend to produce small quantities of deep-coloured wines with relatively little alcohol. Used mainly for rosés, but some Callet has shown it can make a lively young red.

🍇 **CAMARALET,** obscure South Western French variety which can give strongly flavoured wine but is almost extinct.

CAMARÈSE, synonym in the village of Chusclan for the southern Rhône rarity VACCARÈSE, according to Livingstone-Learmonth.

🍇 **CANAIOLO** or **CANAIOLO NERO,** grown all over central Italy and, perhaps most famously, a softening ingredient in the controversial recipe for Chianti, in which it played a more important part than SANGIOVESE in the 18th century. It has declined dramatically in popularity and it does not graft easily. It needs low yields and a reasonably warm autumn to ripen fully but can make attractive, early-maturing wines. Invariably treated as a minor blending ingredient for Sangiovese. Canaiolo is also grown, to more limited extent, in Latium, Sardinia, and the Marches. Italy's total plantings of Canaiolo Nero declined from 6,600 ha/16,300 acres to 4,300 ha in the 1980s, although the variety is allowed in 17 different DOC wines.

A light-berried **Canaiolo Bianco** is also grown in Umbria, where, in Orvieto, it is known as Drupeggio, but it has been declining in popularity.

🍇 **CANNONAU,** sometimes spelt **CANNONAO,** Sardinian (and therefore original, according to Sardinians) name for the widely planted variety known in Spain as Garnacha and in France as GRENACHE. Grown mainly on the east of the island to produce a varietal Cannonau di Sardegna, which comes in several forms, but most commonly as a full throttle dryish red. Although being pulled out by some growers, Cannonau was still grown on 11,000 ha/27,000 acres of the island in 1990.

CAÑOCAZO, See FALSE PEDRO.

CAPE RIESLING, old South African name for CROUCHEN.

🍇 CARIGNAN, known as **Carignane** in the USA, **Carignano** in Italy, and **Cariñena** in Spain, is quantitatively extremely important and qualitatively fairly disastrous. France's most planted variety, five times more common than fashionable CHARDONNAY for example, still dominates the vineyards of the Languedoc, despite copious bribes to pull it out and reduce the European wine surplus. It is better than the ARAMON it replaced in the 1960s to fill the void left in the national blending vat by the independence of Algeria, but only just.

From the perspective of the 1990s, Carignan seems a very odd choice indeed, although presumably it seemed obvious to many *pieds noirs* returning from Algeria, where the wine industry depended at one time on its 140,000 ha/346,000 acres of Carignan. (Carignan still plays a significant part in North African wine production.) Its wine is high in everything—acidity, tannins, colour, bitterness—but flavour and charm. This gives it the double inconvenience of being unsuitable for early consumption yet unworthy of maturation. The vine is not even particularly easy to grow, being extremely sensitive to powdery mildew, quite sensitive to downy mildew, prone to rot and prey to infestation by grape worms. Its diffusion has been extremely beneficial to the agro-chemical industry. Its bunches keep such a tenacious hold on the vine that it does not adapt well to mechanical harvesting, but then the majority of Carignan is not trained on wires anyway but grows in gnarled old bushes. Carignan's great attraction, in the thirsty mid 20th century anyway, was its productivity, up to 200 hl/ha (11.4 tons/acre). It also buds late, which gave it extra allure as a substitute for the much lighter Aramon, previously France's number one vine, which had been badly affected by the frosts of 1956 and 1963. It ripens late too, however, which limits its cultivation to Mediterranean wine regions. In the early 1990s, despite energetic pulling out of Carignan's tenacious stumps all over the Languedoc-Roussillon, there were still 140,000 ha/345,900 acres of these ageing vines.

The regulations for the Languedoc-Roussillon's appellations have been forced to embrace the ubiquitous Carignan, which still accounts for almost half of all vines planted there. But it is hard to argue that, for example, Minervois or Corbières are improved by their (continually reduced) Carignan component. Those wines that depend most heavily on the 'improving' varieties such as Syrah and Mourvèdre and least on Carignan are almost invariably the most successful.

Only the most carefully farmed old vines on well-placed, low-yielding sites can produce Carignan with real character.

Elsewhere, the widespread introduction of carbonic maceration has helped disguise, if not exactly compensate for, Carignan's lack of youthful charm. The astringency of France's basic vin de table has owed much to this vine, although blending with Cinsaut or Grenache helps considerably.

Although the vine almost certainly originated in Spain in the province of Aragon, it is not widely planted there today (although its high acid level is more prized there). Carignan is not even the principal grape variety in the wine that carries its Spanish synonym Cariñena. It is grown chiefly in Catalonia today, although it was historically, as Mazuelo or Mazuela, a not particularly distinguished ingredient in Rioja. It also plays a part in the wines of Costers del Segre, Penedès, Tarragona, and Terra Alta, so that Spain had total plantings of around 12,000 ha/29,600 acres in the early 1990s. Synonyms include Samsó and Crujillon.

The vine, gaining a vowel as Carignane, has been important in the Americas. Although rarely seen today as a varietal, there were still nearly 9,000 acres/3,600 ha in California's hotter regions in 1994 for the vine's productivity and vigour are valued by growers. It is also grown to a considerable extent in Mexico and (to a much lesser extent) in Argentina, Chile, and Uruguay.

Because of its late ripening habits, Carignan can thrive only in relatively hot climates. At one time it underpinned Israel's wine industry and it is by no means unknown in Italy. As Carignano it is grown in Latium and quite successfully in Sardinia, notably in some lively, toothsome Carignano del Sulcis, perhaps as a result of that island's long dominance by Aragon, where it makes strong, ripe reds and rosés. Italy's total plantings of Carignano had fallen to about 2,500 ha by 1990.

The white mutation **Carignan Blanc** can still be found in some vineyards of the Languedoc and, in particular, Roussillon.

CARINA. See CURRANT.

🍇 **CARMENÈRE,** rare old Bordeaux variety which was widely cultivated in the Médoc in the early 18th century and, with Cabernet Franc, established the reputations of its best properties. Daurel reports that the vine is vigorous and used to produce exceptionally good wine but was abandoned because of its susceptibility to coulure and resultant low yields. Its name may well be related to the word 'carmine' and even today it yields small quantities of exceptionally deep-coloured, full bodied wines and may even be, like PETIT VERDOT, the subject of a revival, notably by Carmen of Chile, who have decided to sell their varietal version under the synonym Grande Vidure.

CARMENET, a Médocain synonym for CABERNET FRANC, has also been adopted as the name of a winery in northern California.

❦ **CARNELIAN,** crossing of an earlier (1936) crossing of Carignan and Cabernet Sauvignon with Grenache developed specifically for California conditions by Dr H. P. Olmo and released by the University of Davis in 1972. It was supposed to be a hot climate Cabernet but too many of the Grenache characteristics predominate to make it easy to pick. The liberal produce of its 1,000 acres/400 ha in the Central Valley goes into blends. Curiously, one of the better examples has come from a Texas vineyard, Fall Creek.

❦ **CASTELÃO FRANCÊS,** widely planted all over southern Portugal and known variously as Periquita in Arrábida, Palmela, and Ribatejo; as João de Santarém or Santarém in parts of Ribatejo; Mortágua in the Oeste; and possibly even Trincadeira Preta in parts of the country. This versatile vine can produce fruity, relatively fleshy red wines, sometimes with a slightly gamey character, which can be drunk young or aged.

❦ **CASTETS,** almost extinct vine probably selected from the wild in the harsh Aveyron of South West France.

❦ **CATARRATTO,** the dominant Sicilian variety and, thereby, second most planted single light-berried variety (after all the various TREBBIANOS) in all of Italy in the 1990 agricultural census. According to official returns, there were about 65,000 ha/160,000 acres of **Catarratto Bianco Comune** and nearly 10,000 ha of **Catarratto Bianco Lucido**, the latter, superior in terms of wine quality, having given up ground to the former during the 1980s. Planted almost exclusively in the far western province of Trapani, and specified in the regulations of just three DOC zones, Catarratto was much used for Marsala but today it plays a major role in Italy's (and indeed Europe's) wine surplus. So impressive are Sicily's vineyard dimensions that Catarratto is one of the world's six most planted light-berried wine grapes. Wines have good acid but yields have to be controlled to achieve an attractive aroma.

❦ **CATAWBA,** deep pink-skinned American hybrid grown to a great extent in New York State, especially around the Finger Lakes, where the warmer climate mitigates its tendency to ripen late. This variety, probably a *labrusca × vinifera* hybrid, was identified in North Carolina in 1802, even before CONCORD. It can produce slightly foxy wines of all shades of pink, ingredients in sparkling wines, and can even yield light reds.

❦ **CAYETANA,** high-yielding Spanish vine producing neutral flavoured wine. It is particularly popular in the Extremadura region in the south west, much of its produce being distilled into brandy de Jerez. In Rioja it may be known as Cazagal.

CAZAGAL, see CAYETANA.

CENCIBEL

CENCIBEL, synonym for the Spanish black grape variety TEMPRANILLO, especially in central and southern Spain, notably in La Mancha and Valdepeñas where it is the principal dark-skinned variety.

CENTURIAN, California crossing with same parentage as CARNELIAN but released three years later in 1975. The state's total acreage was static throughout the 1980s at just over 500 acres, all in the Central Valley. It has viticultural advantages over Carnelian but no organoleptic distinction.

CERCEAL, name of several white Portuguese grape varieties, whose Anglicized form is SERCIAL. **Cerceal do Dão,** an ingredient in the heavy white wines of the Dão region, is distinct from the Sercial with which Madeira is associated. Forms of Cerceal are planted in many mainland wine regions where, because of the grapes' high acidity, they are often called ESGANA CÃO, or 'dog strangler'.

CEREZA, pink-skinned, extremely prolific grape variety, which takes its name from the Spanish for cherry and is of increasingly historical interest in Argentina. Like the even more widely planted CRIOLLA GRANDE, it is thought to be descended from the seeds of grapes imported by the early Spanish settlers, although its loose bunches have larger berries which result in paler wine than Criolla. It is declining in importance but there were still about 40,000 ha/100,000 acres of it at the end of the 1980s, most notably in San Juan province. It produces deep white and rosé wine of extremely mediocre quality for early, domestic consumption.

CESANESE, historic vine apparently indigenous to the vineyards surrounding Rome. The superior **Cesanese d'Affile** is more common but is losing ground and was planted on barely 1,300 ha/3,200 acres of vineyard in 1990. Its potential seem barely realized today. **Cesanese Comune** has larger berries and is also known as Bonvino Nero.

CÉSAR, speciality of the far north of Burgundy, notably Irancy, where it may make some contribution to light, soft reds.

CHAMBOURCIN, a relatively recent French hybrid commercially available only since 1963 and popular in the 1970s, particularly in the Muscadet region where, thanks to its resistance to fungal diseases, it was still the third most planted variety according to the 1988 French agricultural census, which found a total of 1,200 ha/3,000 acres. This extremely vigorous, productive vine produces better quality wine than most hybrids, being deep-coloured, aromatic and untainted by foxiness, although it is not officially allowed even into the local vins de pays. It has also been grown on an experimental basis in New South Wales in Australia in a culture unfettered by anti-hybrid prejudice.

☙ **CHANCELLOR,** productive French hybrid developed from two Seibel parents. For long it was known as Seibel 7053, but was named Chancellor in New York in 1970. See SEIBEL for more details.

☙ **CHARBONO,** California's obscure speciality may be the DOLCETTO of Italy, according to Galet, or Argentina's BONARDA, according to Alcalde. **Charbonneau** is a synonym of the virtually extinct Douce Noire of the Savoie region in the French alps, which Galet argues is one and the same as Dolcetto Nero (as the name certainly suggests, although some Italian authorities refute this). The California version called Charbono clings to existence on a handful of acres in the Napa Valley where, partly thanks to the age of the vines, it can produce characterful concentrated reds.

☙ **CHARDONNAY,** a name now so familiar to modern wine lovers around the world that many do not realize that it is the name of a vine variety. In its Burgundian homeland, Chardonnay was for long the sole vine responsible for all of the finest white burgundy. As such, in a region devoted to geographical labelling, its name was known only to vine-growers. All this changed with the advent of varietal labelling in the late 20th century, when Chardonnay virtually became a brand name.

So popular is it that synonyms are rarely used (although some Austrians in Styria persist with their name Morillon). The wine's relatively high level of alcohol, which can often taste slightly sweet, has probably played a part in this popularity, as has the obvious appeal of the oak so often used in making Chardonnay.

But it is not just wine drinkers who appreciate the broad, easy-to-appreciate if difficult-to-describe charms of golden Chardonnay. (The Australian Wine Research Institute's initiative, analysing the component parts of each major variety's flavour, found Chardonnay a particularly nebulous target, identifying flavour compounds also found in, among other things, raspberries, vanilla, 'tropical fruits', peaches, tomatoes, tobacco, tea, and rose petals.) Vine-growers appreciate the ease with which, in a wide range of climates, they can coax relatively high yields from this vine (whose natural vigour may need to be curbed by either dense planting or canopy management). Wine quality is severely prejudiced, however, at yields above 80 hl/ha (4.5 tons/acre) and yields of 30 hl/ha or less are needed for seriously fine wine. Growers' only major reservation is that it buds quite early, just after Pinot Noir, which regularly puts the coolest vineyards of Chablis and Champagne at risk from spring frosts. It can suffer from coulure and occasionally millerandage and the grapes' relatively thin skins can encourage rot if there is rain at harvest time, but it can thrive in climates as diverse as those of Chablis in northern France and Australia's hot Riverland. Picking time is crucial for, unlike Cabernet Sauvignon, Chardonnay can lose its crucial acidity fast in the latter stages of ripening.

Wine-makers love Chardonnay for its reliably high ripeness levels and its malleability. It will happily respond to a far wider range of wine-making techniques than most white varieties. The Mosel or Vouvray wine-making recipe of a long, cool fermentation followed by early bottling can be applied to Chardonnay. Or it can be fermented and aged in small oak barrels, some of the highest-quality fruit being able to stand up to new oak. It accommodates each individual wine-maker's policy on the second, softening malolactic fermentation and lees stirring without demur. Chardonnay is also a crucial ingredient in most of the world's best sparkling wine, not just in Champagne, demonstrating its ability to age in bottle even when picked early. And, picked late, it has even been known to produce some creditable sweet wines, notably in the Mâconnais, Romania, and New Zealand, from grapes attacked by noble rot.

Chardonnay also manages to retain a remarkable amount of its own character even when blended with other less fashionable varieties such as Chenin Blanc, Sémillon, or Colombard to meet demand at the lower end of the market. But perhaps this is because its own character is, unlike that of the other ultra-fashionable white, Sauvignon Blanc, not too pronounced. Chardonnay from young or over-productive vines can taste almost aqueous. Basic Chardonnay may be vaguely fruity (apples or melons) but at its best Chardonnay, like Pinot Noir, is merely a vehicle for the character of the vineyard in which it is grown. In many other ambitious wines fashioned in the image of top white burgundy its 'flavour' is actually that of the oak in which it was matured, or the relics of the wine-making techniques used (see above). When the vineyard site is right, yields are not too high, acid not too low, and wine-making skilled, Chardonnay can produce wines that will continue to improve in bottle for one, two, or, exceptionally, more decades but—unlike RIESLING and the best, nobly rotten CHENIN BLANC and SÉMILLON—it is not a variety capable of making whites for the very long term.

Chardonnay's origins are obscure. For long it was thought to be a white mutation of Pinot Noir, and it has often been called Pinot Chardonnay, but Galet cites good ampelographical evidence for Chardonnay's being a variety in its own right. A village in the Mâconnais called Chardonnay has excited various theories, while others speculate that Chardonnay's origins may be Middle Eastern, evincing its long history in the vineyards of Lebanon.

There is a rare but distinct pink-berried mutation, **Chardonnay Rose**, as well as a headily perfumed **Chardonnay Blanc Musqué** version, sometimes used in blends. Some of the 34 official French clones of Chardonnay have a similarly grapey perfume, notably 77 and 809, which are now quite widely disseminated and can add a rather incongruously aromatic note to blends with other clones of the variety. The arguably over-enthusiastic application of clonal selection techniques in Burgundy means that growers can now choose from a wide range of Chardonnay clones specially selected for their productivity, particularly 75, 78, 121, 124, 125, and 277. Those seeking quality rather than quantity are more likely to choose 76, 95, and 96.

Chardonnay cuttings are sought after the world over and in many countries—France, America, even fiercely quarantined Australia, New Zealand, and South Africa—this is the white variety for which nurserymen found the greatest demand in the late 1980s, fuelled partly by consumer demand for full bodied white wines with the magic word Chardonnay on the label, but also by the dramatic expansion of the world's sparkling wine industry. Such is Chardonnay's glamour that it has probably been the variety of which most cuttings have been smuggled by ambitious wine producers thwarted by plant quarantine regulations.

Although in terms of total area planted, this superior quality white variety is dwarfed by those traditionally used for brandy production such as AIRÉN and TREBBIANO, its popularity in the late 1980s was sufficient to propel it to first or second place in terms of area planted in each of France, California, Washington State, Australia, and New Zealand in the early 1990s. It thereby overtook Sémillon and Riesling in the 1980s to occupy more of the world's vineyard than any white-berried variety other than Airén, Trebbiano, RKATSITELI and possibly Sicily's CATARRATTO—a remarkable feat for a vine credited with such nobility.

In France, for example, Chardonnay plantings virtually doubled from 1980 to reach 25,000 ha/62,000 acres by 1993. This was initially due to the expansion of the Champagne vineyard, where Chardonnay now represents as much as a third of all vines planted, and of the Chablis zone (where the grape can achieve a lean steeliness and, in the best examples, considerable longevity). Improvements in frost protection techniques played a part in both these expansions.

In the Burgundian heartland, the Côte d'Or, Chardonnay plantings increased by one-quarter in the 1980s to a grand total of 1,400 ha—just half as much Chardonnay as was actually planted in a single year, 1988, in California! Perhaps it is not so surprising that Meursault is a name unknown to legions of Chardonnay drinkers. Although Chardonnay, sometimes called Beaunois or Aubaine in Burgundy, has gradually been replacing GAMAY and ALIGOTÉ, Pinot Noir vines still outnumber Chardonnay vines more than four to one on the Côte d'Or. Notably more Chardonnay is grown on the southern Côte de Beaune than the Côte de Nuits. Famous white wine appellations with typical characteristics in brackets include, from north to south, Corton-Charlemagne (marzipan), Meursault (buttery), Puligny-Montrachet (fine and steely), Chassagne-Montrachet (hazelnuts), and any name that includes the word Montrachet (enormous concentration, and alcohol levels of 13 per cent and above).

In the Côte Chalonnaise and Mâconnais to the south, plantings overtook those of Gamay in the 1980s so that there were 4,500 ha by 1988. From the Côte Chalonnaise, the whites of Rully, Mercurey, and Montagny can offer economic, if slightly rustic versions of the grander names of the Côte d'Or. The Mâconnais, where Chardonnay can take on a broad, appley character, produces not just white Mâcon with a range of geographical suffixes but also various Pouillys, most famously the full bodied Pouilly-Fuissé. From further south still come the very similar St-Véran and Beaujolais Blanc. Although the regulations allow Aligoté into Beaujolais Blanc and PINOT BLANC into white wines labelled Bourgogne and Mâcon, most of these less expensive white burgundies are in practice made predominantly from Chardonnay. To the horror of the appellation authorities, there is an increasing trend towards slipping the word Chardonnay on to white burgundy labels to increase their appeal to non-French consumers.

Although nearly three-quarters of France's Chardonnay is still in either Champagne or Greater Burgundy, the variety has been sweeping south and west from this base. It is embraced by an ever wider variety of appellations and plantings can be found in Alsace, Ardèche, Jura, Savoie, Loire, and, especially, the Languedoc, where it was first planted to add international appeal to the lemony wines of Limoux. Official figures suggest that a total of more than 5,200 ha/13,000 acres was planted in the Languedoc between 1988 and 1993 and much of it is now used in varietal Vins de Pays d'Oc of extremely varying quality.

Few would have believed in 1980, when California had just 18,000 acres/7,200 ha of Chardonnay, that by 1988 the state's total plantings would overtake the (rapidly increasing) French total so that by 1994 California had 66,590 acres. The rate of new plantings reached a peak in 1988, however, and the early 1990s brought a new red wine fashion for California grape-growers to grapple with. Nearly half of all California Chardonnay is concentrated in Sonoma, Napa, and Monterey counties but there are also sizeable plantings further south in Santa Barbara and San Luis Obispo. Quality varies, from ambitiously priced wines which are noticeably more consistent and user-friendly than the Burgundian classics to sweet commercial blends, but the archetype remains much the same: glossy, golden wines with a 'kiss of oak'.

Chardonnay, now North American for 'white wine', has been embraced with equal fervour throughout the rest of North America, from British Columbia to Long Island, New York. In 1990 it overtook Riesling to become the most planted variety of any hue in Washington State with 2,600 acres/1,000 ha and it is also popular, if not always desperately successful, in Oregon and Texas. The scale of America's romance with Chardonnay in general and oak-aged Chardonnay in particular has had a profound effect on the structure of the international cooperage business.

Various South American countries have been seeking out cooler spots to imbue their Chardonnay with real concentration. Chile's Casablanca region and Argentina's Tupungato are the most obvious examples, whose best wines combine those New World virtues of accessibility and value.

The Australian wine industry's all-important export trade has been centred on its peculiarly exuberant style of Chardonnay. Rich fruit flavours, often disciplined by added acid and flavoured by oak chips, are available at carefully judged prices. Such was the strength of demand for Australian Chardonnay in the late 1980s that the area of Chardonnay vines increased more than fivefold during the decade so that in 1990 Chardonnay, with its 4,300 ha/10,600 acres, became Australia's most planted white wine grape variety (although 1,300 ha was too young to bear fruit). Wines vary from limey essences grown in cooler spots in Victoria and Tasmania to almost syrupy, smokey blends concocted from the hot irrigated vineyards of the interior. The average life expectancy of an Australian (and most other New World) Chardonnay is short.

Nor has New Zealand escaped Chardonnay-mania and by 1992 only the more traditional MÜLLER-THURGAU covered more ground. New Zealand's Chardonnays have perceptibly more natural acid than their trans-Tasman neighbours.

Chardonnay has had a chequered history in South Africa, with some of the first cuttings so named being eventually identified as the very much less fashionably exciting AUXERROIS. The quality of the clones planted has since improved steadily.

Although Chardonnay can thrive in relatively hot climates (such as Australia's irrigation zones), it has to be picked before acids plummet (often before the grapes have developed much real character) and it does require relatively sophisticated techniques, and access to cooling equipment, in the cellar. This is why it is not well suited to the less developed Mediterranean wine regions. Even in the Lebanon, where local strains of Chardonnay are well entrenched, the wine tends to betray its torrid origins. With human skill and investment in technology, however, more well-balanced Chardonnays with the real interest of isolated examples from Israel may emerge.

The variety continues to be planted in an ever wider range of locations, but Italy has a long history of Chardonnay cultivation, especially on its subalpine slopes in the north. For decades Italians were casual about distinguishing between their Pinot Bianco (PINOT BLANC, also known as Weissburgunder in the Italian Tyrol) and their Chardonnay (traditionally called Gelber, or Golden, Weissburgunder in the Italian Tyrol). Indeed the Italian agricultural census of 1982 failed to distinguish a single Chardonnay vine, while that of 1990 located more than 6,000 ha/15,000 acres.

International market forces eventually convinced Italians that the distinction could be worthwhile, although the Italian authorities were slow to recognize this dangerously Gallic grape as a name officially allowed into any DOC. Alto Adige Chardonnay was the first accorded DOC status, in 1984, although the vine has since been working its magic on producers all over Italy from Apulia to Piedmont and, of course, Aosta towards the

French border. Nowadays much of Italy's Chardonnay is produced, often without much distinction, in Friuli, Trentino, and to a more limited extent the Veneto, where much of it is used as ballast for Soave's GARGANEGA. Some fine examples are produced in favoured sites in both Friuli and Trentino, but a considerable proportion is siphoned off for sparkling wine. Most of Italy's most ambitious still Chardonnays are fairly well oaked, exhibiting every possible wine-making technique. The vine has been gaining ground rapidly in Italy, being planted in Tuscan spots where Sangiovese is difficult to ripen and in Piedmont replacing Dolcetto, which can be difficult to sell.

Much less dramatic Chardonnay is also produced in Switzerland, particularly in Geneva and Valais. In Austria, a foreign vine known as Morillon in Styria and Feinburgunder in Vienna and Burgenland was not identified as the modish Chardonnay until the late 1980s (when Styria had well over 200 ha of it). Austria's Chardonnays include relatively rich, oak-matured versions; lean, aromatic styles modelled on their finest Rieslings; and even sweet Ausbruch wines.

Bulgaria has a vast area of Chardonnay vineyard but, perhaps because of over-production or for wine-making reasons, is rarely able to demonstrate real Chardonnay character in the bottle. There are limited plantings in Slovenia, Hungary, and Romania (whence Late Harvest Chardonnays have been exported) but it seems that the Soviet Union's political turbulence during the late 1980s may have saved it from the major Chardonnay invasion that took place almost everywhere else at that time. Official statistics in 1993 found that it had infiltrated only Moldova and Georgia, and played an extremely minor role relative to, for example, the white grape varieties RKATSITELI, RIESLING, and Chardonnay's Burgundian rival ALIGOTÉ.

Germany was one of the last wine-producing countries to admit Chardonnay to the ranks of accepted vine varieties, in 1991, to a very limited extent, which is perhaps not surprising since giving over one of Germany's favoured sites to this quintessentially French variety is inevitably viewed by some as a defeat for Germany's signal white variety Riesling.

Such is Chardonnay's fame and popularity that it is grown to a certain extent in climates as dissimilar as those of England, India, and Uruguay. In Spanish Catalonia Chardonnay has added class and an internationally recognizable flavour to Cava sparkling wines as well as producing some relatively fat still wines both here and in Costers del Segre and Somontano. Portugal, almost alone, seems to have withstood Chardonnay-mania.

🍇 **CHASAN,** southern French crossing of PALOMINO (known in France as Listan) and Chardonnay made by French ampelographer Paul Truel. The resulting wine, whose name usefully, in commercial terms, begins with the same three letters as fashionable Chardonnay, bears a lightweight imprint of Chardonnay flavour and structure and the vine buds early. It is planted on a limited scale in the Languedoc, particularly in the Aude.

🍇 **CHASSELAS,** vine with a long, intriguing history that is widely planted around the world. Some authorities cite Middle Eastern, even Egyptian origins. Others cite France and the Swiss, who undoubtedly make the finest examples today, claim it for their own, pointing out that the name Fendant, its common synonym in the Valais, can be found in monastic records well before the 16th century.

In France it is rather despised, not least because, as Chasselas Doré or Golden Chasselas, it is France's most common table grape. It is rapidly disappearing from Alsace, where it is regarded as the lowest of the low and is generally sold as Edelzwicker or under some proprietary name that excludes mention of any grape variety. Planted in the Pouilly-Fumé zone, it makes the distinctly inferior white labelled Pouilly-sur-Loire. French Chasselas is at its most noble in Crépy in alpine Savoie, where it has a long history.

Here on the shores of Lake Geneva, as on the other side of the lake in Switzerland, care has to be taken with the choice of rootstock so as to avoid the variety's dangerous tendencies towards early budding and too many leaves. But skilfully grown Chasselas can yield good quantities of fairly neutral, soft wine which achieves a peak of concentration in isolated sites such as some round the village of Dézaley. Its Vaud name is Dorin and overall Chasselas is by far Switzerland's most planted variety.

The variety's long history has enabled it to spread far and wide. In Germany, where it is known as Weisser Gutedel, it has also been known since the 16th century. It was once revered in the Pfalz region and there were still more than 1,300 ha/3,200 acres planted in Germany in 1990. In Austria it is known, but not widely grown, as Moster and Wälscher. It is grown widely in Romania (mainly for the table), in Hungary (where it is the second most planted grape), to a limited extent in Moldova and Ukraine, in both the north and far south of Italy (where it is sometimes called Marzemina Bianca), around the Mediterranean including North Africa, in Chile, and was at one time curiously important in New Zealand.

The variety known for some time as Golden Chasselas in California is thought to be PALOMINO.

🍇 **CHENEL,** South African crossing of CHENIN BLANC × TREBBIANO developed for its resistance to rot. See also WELDRA.

🍇 **CHENIN** or **CHENIN BLANC,** in its native region often called Pineau or Pineau de la Loire, is probably the world's most versatile grape variety, capable of producing some of the finest, longest-living sweet whites, although more usually harnessed to the yoke of basic New World table wine production. In between these two extremes it is responsible for a considerable volume of sparkling wine and in South Africa, where it is by far the most planted vine, it is even used as the base for a wide range of fortified wines and spirits. Although in its high-yield, New World form its distinctive flavour reminiscent of honey and damp straw is usually lost, it retains the naturally high acidity that dogs it in some of the Loire's less ripe vintages but can be so useful in a hot climate.

South Africa now has more than three times as much Chenin planted as France and the Cape's strain of the variety, often called Steen, constitutes nearly 30 per cent of the country's entire vineyard and was almost certainly one of the original collection imported in 1655 by Jan Van Rieebeck. Most of its South African produce is off-dry, refreshingly crisp but otherwise rather bland, but there are increasing experiments with blending, and making sweet and even oaked versions.

California also has more Chenin planted than France, and uses it for much the same purposes as South Africa, as the often anonymous base for everyday commercial blends of reasonably crisp white of varying degrees of sweetness, often blended with the even more widely planted French COLOMBARD. Both of these workhorse varieties are planted primarily in the hot Central Valley, a setting that might be described as the antithesis of Chenin's Loire homeland (although Clarksburg can produce some intriguing melony, musky flavours). Only a handful of producers take Chenin seriously enough to try to make wines worth ageing from it. One or two try to make a wine in the image of the great sweet Loire Chenins, as do one or two producers in New Zealand where there were 200 ha/500 acres in the early 1990s, mainly in the North Island. Australia had three times as much, much of it misidentified in its time, but treats it largely with disdain as suitably acid blending material, usually extending Chardonnay, or even spiking a blend of Chardonnay and Semillon.

Chenin is widely planted throughout the Americas for no perceptible reason other than that it will obligingly produce a decent yield of relatively crisp wine. There are perhaps as many as 4,000 ha/10,000 acres of Chenin in Argentina, much of it used for early maturing sparkling wines, as well as substantial plantings in Chile. In Mexico, Brazil, and Uruguay it is still more usually called Pinot Blanco. It is also common, though not particularly popular, in many North American states outside California. In Washington, for example, almost half of all Chenin vines were pulled out in the late 1980s. The variety was also exported to Israel to establish vineyards there at the end of the 19th century.

If Chenin appears to lead a double life—biddable workhorse in the New World, superstar in Anjou-Touraine—it seems clear that the explanation lies in a combination of climate, soil, and yield. In California's Central Valley the vine is often expected to yield 10 tons per acre (175 hl/ha), while even the most basic Anjou Blanc should not be produced from vines that yield more than 45 hl/ha. It is hardly surprising that Chenin's character seems diluted outside the Loire. Galet suggests that it may have been well established in Anjou as long ago as the ninth century and that it was exported to Touraine in the 15th. Rabelais certainly wrote about Chenin both as Chenin and its already familiar synonym of Pineau, often Pineau d'Anjou.

The vine is vigorous and has a tendency to bud early and ripen late, both of which are highly inconvenient attributes in the cool Loire Valley (though hardly noticeable characteristics in the hotter vineyards of the New World). Clones that minimize these inconveniences have been selected and six had been officially sanctioned in France by the 1990s.

About a third of all France's, which means the middle Loire's, Chenin was abandoned in the 1970s, often in favour of the red Cabernet Franc in Anjou-Touraine and to make way for the then more fashionable Gamay and Sauvignon de Touraine in the east of the middle Loire. It is today the second most planted variety in the heart of Anjou-Touraine, as well it might be to judge from the superlative quality of the best wines of such appellations as Anjou, Bonnezeaux, Coteaux de l'Aubance, Coteaux du Layon, Jasnières, Montlouis, Quarts de Chaume, Saumur, Savennières, Vouvray, and Crémant de Loire.

In most of the best wines, and certainly all of the great sweet wines, Chenin is unblended, but up to 20 per cent of Chardonnay or Sauvignon is allowed into an Anjou or a Saumur and even more catholic blends are allowed into whites labelled Touraine— although even here Chardonnay's pervasive influence is officially limited to 20 per cent of the total blend. If Loire white has any character at all it is that of Chenin, however traduced in the rest of the world.

While basic Loire Chenin exhibits simply vaguely floral aromas and refreshingly high acidity (together with too much sulphur if made in one of the more old-fashioned cellars), the best has a physically thrilling concentration of honeyed flavour, whether the wine is made sweet (*moelleux*), dry, or demi-sec, together with Chenin's characteristically vibrant acidity level. In classic years noble rot helps preserve the finest Chenins for decades after their relatively early bottling. (In all of these respects, together with lateness of ripening and a wide range of sweetness levels that are customary, Chenin is France's answer to Germany's Riesling.)

Chenin with its high acidity is a useful base for a wide range of sparkling wines, most importantly Saumur Mousseux but also Crémant de Loire and even some rich sparkling Vouvrays which, like their still counterparts, can age beautifully. Treasured for its reliably high acidity, and useful perfume, it is also an ingredient, with Mauzac and, increasingly, Chardonnay in fizzy Limoux.

The twin great vintages for middle Loire sweet whites of 1989 and 1990 did something to raise the profile, and price, of great Chenin, but a variety whose best wines demand time to be fully appreciated inevitably suffers in the late 20th century.

Chenin Noir is a rarely used synonym for PINEAU D'AUNIS, a dark-berried grape variety.

CHEVRIER, historic name for SÉMILLON, occasionally revived.

CHIAVENNASCA, synonym for the noble NEBBIOLO vine and grape in Valtellina.

🍇 **CILIEGIOLO,** central Italian red grape variety of Tuscan origin named after its supposed cherry-like flavour and colour. It is declining in popularity and total Italian plantings of the vine were just 5,000 ha / 12,400 acres in 1990, although it can make some excellent wines, and could be a usefully soft blending partner for SANGIOVESE.

🍇 **CINSAUT,** often written **Cinsault,** rather GRENACHE-like grape variety known for centuries in the Languedoc region of southern France. Although it has good drought resistance, it can all too easily be persuaded to yield generously and unremarkably and its best wines, with a heady aroma not unattractively reminiscent of paint, come from vines that yield less than 40 hl/ha (2.3 tons/acre). The wines it produces tend to be lighter, softer, and, in extreme youth, more aromatic than most reds. Although prone to rot, it is particularly well adapted for rosé production and is widely planted throughout southern France, and Corsica where it is the dominant vine variety. It ripens earlier than Grenache and its softer wood makes it more suitable for mechanical harvesting.

There was a threefold increase in French plantings in the 1970s when Cinsaut was officially sanctioned as an 'improving' grape variety with which to replace ARAMON and ALICANTE BOUSCHET in the Languedoc. Since then, the economic realities of quality's supremacy over quantity have slowed Cinsaut's fortunes. Cinsaut is used almost exclusively to add suppleness, perfume, and immediate fruit to blends (typically of the ubiquitous but curmudgeonly CARIGNAN), although some all-Cinsaut rosés exist.

In the Rhône it is an approved but hardly venerated ingredient in the Châteauneuf-du-Pape cocktail and is often found further east in Provence, as well as in the north of Corsica, where it has been widely pulled up in favour of more profitable crops.

Total French plantings of Cinsaut fell throughout the 1980s to less than 50,000 ha / 123,500 acres (still more vineyard than Cabernet Sauvignon). The variety was most important in the 1950s and early 1960s when Algeria, then constitutionally part of France, was an important wine producer and depended particularly heavily on its healthily productive 60,000 ha of Cinsaut. Since Algerian wine was then used primarily for blending in France, notoriously for adding body to less reputable burgundies, some of this North African Cinsaut may still be found in a few older bottles of 'burgundy'. The variety is still the most planted wine grape in Morocco and has long played a part in the wine industries of Tunisia, Algeria, and Lebanon.

Cinsaut is still important in South as well as North Africa. Having been imported from southern France in the mid 19th century, it was South Africa's most planted grape variety until the mid 1960s and was only recently overtaken by CABERNET SAUVIGNON to become the Cape's most popular red grape variety. South Africa had 4,200 ha/10,400 acres of Cinsaut planted in 1994. It was known carelessly as Hermitage in South Africa (although there is no Cinsaut in the northern Rhône). Thus South Africa's own grape variety speciality, a crossing of Pinot Noir with Cinsaut, was named PINOTAGE, now a much more respected South African vine variety than Cinsaut.

In both France and Australia (where its fortunes waned rapidly in the 1970s and 1980s) Cinsaut has occasionally been sold as a table grape under the name Oeillade. In southern Italy, it is probably the same as the Ottavianello, planted around Brindisi and producing light, unremarkable red wines. Cinsaut can also be found in various corners of eastern Europe.

Ⓦ **CLAIRETTE,** much-used name for a variety of southern French white grape varieties. Clairette Ronde, for example, is the Languedoc name for the ubiquitous UGNI BLANC, and various Clairettes serve as synonyms for the much finer BOURBOULENC.

True **Clairette Blanche**, however, is a decidedly old-fashioned variety, producing slightly flabby, alcoholic whites that oxidize easily, but it is allowed into a wide range of southern Rhône, Provençal, and Languedoc, appellations, even lending its name to three (Bellegarde, Languedoc and the sparkling Die version). Clairette needs a non-vigorous rootstock to avoid coulure, and is susceptible to mildew and grape worms. Its small, thick-skinned grapes can ripen dangerously fast at the end of the growing season. Total French plantings fell by more than half during the 1970s and there were just 4,000 ha/10,000 acres at the end of the 1980s, a decade during which consumers acquired a taste for whites that in many ways are the antithesis of Clairette. It is still, usually enlivened by Ugni Blanc and Terret, one of the principal ingredients in the Languedoc's white vins de pays.

Its presence in many southern white appellations such as Lirac, Costières de Nîmes, and Palette explains why Ugni Blanc is also needed in these blends, to add counterbalancing acid. Its other common partner in the blending vat, and often vineyard, GRENACHE Blanc certainly does nothing to offset Clairette's weight and premature senility, although low-temperature fermentation and minimal exposure to oxidation can do something to offset these tendencies.

In previous eras when consumers expected their whites to look pale brown and taste halfway to sherry, Clairette was clearly relatively important. With the more acid PICPOUL it formed the basis of picardan, an extraordinarily popular wine exported in enormous quantities northwards from the Languedoc in the 17th and 18th centuries. It is hardly surprising that the variety spread far and wide in the 19th and early 20th centuries. At one time there were sizeable plantings in Algeria. In South Africa it is grown on a relatively important area, at 3,500 ha almost as much as the French total, and provides a useful ingredient for basic sparkling wine or good blending material. It can still be found in Australia's Hunter Valley, where it is known as Blanquette but is declining fast. It is also planted in Romania, Israel, Tuscany, and Sardinia, where it is a permitted ingredient in Nuragus di Cagliari.

A Clairette is grown in Russia in some quantity.

CLARE RIESLING. See CROUCHEN.

🍇 **CLARET DE GERS,** probably the same as **Claret de Gascogne,** an almost extinct and undistinguished South Western French variety called Blanc Dame.

🍇 **CLAVERIE,** South Western French vine, sometimes called Chalosse Blanche, which once produced alcoholic wine sought out by the Dutch but is now, even in the Landes, virtually extinct.

CLEVNER, name usually applied to members of the PINOT family. In Switzerland the name is often applied to PINOT NOIR grown in the canton of Zürich. See also KLEVNER.

🍇 **CLINTON,** *labrusca* American grape with a pronounced foxy taste occasionally found in Italian Switzerland and even, sometimes, across the border in Italy. The vine was planted as a response to the phylloxera invasion of the late 19th century.

🍇 **COCOCCIOLA,** one of the few Italian vines to have expanded its total area, admittedly only to 1,400ha/3,500 acres, between 1982 and 1990. An ingredient in Trebbiano d'Abruzzo.

🍇 **COCUR,** Ukrainian speciality and the republic's fourth most planted wine grape, after ALIGOTÉ, RKATSITELI, and SAUVIGNON BLANC.

🍇 **CODA DI VOLPE,** ancient, full bodied Campanian speciality grown near Naples. Another vine goes by the same name: Emilia.

CÓDEGA, Douro name for ROUPEIRO.

🍇 **COLOMBARD,** originally used with UGNI BLANC and FOLLE BLANCHE to make cognac, but considered inferior to both. As Colombard's star waned in France, almost half of total plantings being pulled up in the 1970s, it waxed quite spectacularly in California where, as French Colombard, it was the state's most planted wine grape variety of all until Chardonnay overtook it in the early 1990s, providing generous quantities of reasonably neutral but reliably crisp base wine for commercial, often quite sweet, white blends.

Its disadvantages of being prone to rot and powdery mildew are much lesser inconveniences in the hot Central Valley, where almost all of California's Colombard is planted (official statistics record just 43 acres/17 ha of the variety ever planted in the smart

Napa Valley). And Colombard's disadvantages for the distillers of Charentes, that its wine is more alcoholic and less acid than that of the other cognac varieties, are positive advantages for consumers of the wine in its undistilled state (although Colombard is also used for California brandy).

It would take some sorcery to transform Colombard into an exciting wine, but pleasantly lively innocuousness is well within reach for those equipped with stainless steel and temperature control. In a nice example of transatlantic switchback, producers in armagnac country set about duplicating California's modern wine-making transformation of the dull Colombard grape on their own varieties surplus to brandy production, thus creating the hugely successful Vin de Pays des Côtes de Gascogne (later copied by their rivals in cognac country). Colombard, still the third most important variety in the region after Ugni Blanc and BACO Blanc, which are particularly well suited to distillation, has been a prime ingredient in this hugely successful wine.

Colombard is still grown all over South West France, and particularly in the north and west of the Bordeaux region around Bourg and Blaye, where much of it acts as a dull subordinate ingredient in dull blended Bordeaux Blanc.

Often called **Colombar**, and once important to the local brandy industry, it also reached a peak of popularity for cheap, commercial off-dry white in South Africa. There were still 5,700 ha / 14,000 acres left in 1990, making it the country's fifth most important variety, and some wine-makers here use more ingenuity than most in attempting to fashion various silk purses from this sow's ear of a vine.

In Australia too there were still 600 ha in the early 1990s, although it is usually blended with more fashionable grape varieties, often adding useful natural acidity.

❦ COLORINO, increasingly rare, thick-skinned, deep-coloured red-fleshed grape variety used traditionally in Tuscany for drying and then adding to deepen Chianti. Less useful with improved SANGIOVESE.

COLUMBARD, occasional name (or misspelling) for COLOMBARD.

❦ COMPLETER, curious speciality of Herrschaft in Graubünden in eastern Switzerland. The wine produced is relatively aromatic and full bodied. According to Sloan this is the same as the Lafnetscha of Valais in French-speaking Switzerland.

❦ CONCORD, native American vine and the most important variety cultivated in the north eastern United States, notably in New York State. It belongs to the species *Vitis labrusca* and the pronounced foxy flavour of its juice makes its wine an acquired taste for those raised on the produce of *vinifera* varieties. It was named after Concord, Massachusetts by Ephraim W. Bull, who introduced it, having planted the seeds of a wild vine there in 1843. It is particularly important for the production of grape juice and grape jelly, but it produces a wide range of wines, often with some considerable residual sugar. Viticulturally, the vine is extremely well adapted to the low temperatures of New York and is both productive and vigorous, if rather late ripening. Some Concord has also been grown in Brazil.

CORDISCO, occasional name for Italy's MONTEPULCIANO.

🍇 **CORNALIN**, waning traditional French Swiss vine which ripens late to produce particularly deep-coloured wine which needs ageing.

🍇 **CORNIFESTO**, minor port grape making light wines.

🍇 **CORTESE**, celebrated in Piedmont, north west Italy for more than a century. Its most highly regarded wine is Gavi, produced initially to serve the fish restaurants of Genoa and the Ligurian coast not far to the south. The Cortese dell'Alto Monferrato a few miles west, like the Cortese grown on the Colli Tortonese, rarely achieves the ripeness, or wine-making proficiency, of Gavi. The wine produced is rarely complex and can be ineffably bland (unlike Piedmont's white ARNEIS and FAVORITA grapes) but sustains a good level of acidity through to full ripeness.

Cortese is also grown in Oltrepò Pavese in Lombardy and may be part of the blend in the Veneto's Bianco di Custoza. Total plantings in Italy fell by a quarter in the 1980s to a total of fewer than 3,000 ha/7,500 acres.

🍇 **CORVINA**, or **CORVINA VERONESE**, the dominant and best grape variety of Valpolicella and Bardolino in north east Italy, producing relatively light, fruity, red wines with a certain almond flavour. Valpolicella DOC regulations stipulating the relatively bland RONDINELLA, and some of the tart MOLINARA, must constitute a combined total of at least 30 per cent of the blend have therefore been criticized. This, sometimes called Cruina, is the variety most prized for drying, as in Amarone and Recioto production. Italy's total plantings of the Corvina Veronese vine variety were nearly 4,500 ha/11,000 acres in the early 1990s.

CÔT or **COT**, important synonym MALBEC, which is known in Cahors as Auxerrois.

🍇 **COUDERC NOIR**, French hybrid of a dark-berried *Vitis rupestris Lincecumii* and *vinifera*, is one of several productive but undistinguished hybrids that proliferated in the south of France in the early 20th century (see BACO, CHAMBOURCIN, PLANTET, SEIBEL, SEYVE-VILLARD, VILLARD). Although not as popular as Villard once was, in terms of total area planted Couderc Noir was not overtaken by Cabernet Sauvignon until well into the 1970s. By this time vigorous steps were being taken to eradicate this embarrassing legacy of another viticultural era, and the 1988 agricultural census found only about 2,500 ha of each of Couderc Noir, Villard Noir, and the various Seyve-Villard varieties. Couderc, which does not even offer particularly good resistance to phylloxera unless grafted on to a vigorous rootstock, needs a hot climate because it ripens so late, too late to be of any use to the eastern states of North America. The wine produced can be aggressively non-*vinifera* in taste.

🌱 **COUNOISE,** one of the more rarefied ingredients in red Châteauneuf-du-Pape, easily confused in the vineyard with the much lesser southern Rhône variety AUBUN, with which it may sometimes be mingled in older vineyards. It is authorized as a supplementary ingredient for most red wine appellations around the southern Rhône, including Coteaux du Languedoc (which allows Aubun as a synonym), but is not widely grown outside Châteauneuf-du-Pape, although total French plantings increased in the 1980s to around 900 ha/2,200 acres.

As a vine, it leafs and ripens late and yields conservatively. As a wine, it is not particularly deeply coloured or alcoholic but adds a peppery note and lively acidity to a blend. Properties such as Château de Beaucastel typically use about 5 per cent of Counoise in their red Châteauneuf-du-Pape.

COURBU BLANC. See PETIT COURBU. The dark-berried **Courbu Noir** version is another Béarn speciality from South West France and is all but extinct.

🌱 **CRATO, CRATO BRANCO,** speciality of Portugal's Algarve.

🌱 **CRIOLLA CHICA,** the Argentine name for a lighter-skinned strain of the PAIS of Chile, the MISSION of California, and the Negra Corriente of Peru. It is thought to be descended from the seeds of grapes, presumably well raisined after their voyage under sail across the Atlantic, imported by the Spanish conquistadores, possibly as early as the 16th century. Although Criolla Chica is much less common in Argentina than the other pink-skinned grape varieties CRIOLLA GRANDE and CEREZA, it is more common in La Rioja province than Criolla Grande.

🌱 **CRIOLLA GRANDE,** Argentina's most planted vine, producing vast quantities of pink or deep-coloured white wine for domestic consumption from its coarse, pink-skinned grapes. Although the area planted is declining, it was nearly 37,000 ha/91,000 acres in 1990, which may have been enough for its inclusion in the world's top 20 vine varieties by area planted. Although it is sometimes called after San Juan province **Criolla Sanjuanina**, Criolla Grande is most widely planted in Mendoza province, where it is the most planted vine variety by far (covering three times the area of the red wine grape MALBEC, for example).

Criolla Grande is a low-quality *vinifera* variety that was probably one of the first vines cultivated in the Americas, and is deeper skinned than CRIOLLA CHICA.

Note: the two Criollas, along with CEREZA and Moscatel Rosada, form the basis of Argentina's declining trade in basic deep-coloured white wine sold very cheaply in litre bottles or cardboard cartons.

🍇 **CROATINA,** red grape variety from the borders of the Piedmont and Lombardy regions of northern Italy. The vine buds and ripens late but yields good quantities of fruity wine with a certain bite, designed to be drunk relatively young. Its common synonym is Bonarda, under which name it is sold as an appetizing varietal DOC in Oltrepò Pavese. The variety is quite distinct from BONARDA Piemontese. Italy's official tally of Croatina plantings declined from 5,500 to 4,500 ha / 11,000 acres in the 1980s.

🍇 **CROUCHEN** or **CRUCHEN,** well-travelled vine responsible for neutral wines in both South Africa and Australia. It originated in the western Pyrenees of France but is no longer grown there in any quantity, thanks to its sensitivity to fungal diseases. There are records of its shipment to Clare Valley in South Australia in 1850 and for long it was confused with Semillon, which Australians were wont to call Riesling. It was therefore known principally as Clare Riesling in Australia until 1976, when ampelographer Paul Truel identified it as this relatively obscure French variety. There were still 420 ha / 1,000 acres planted in Australia in the early 1990s but the wine produced is generally used as gently aromatic blending material. The South Africans had 3,500 ha of the variety they call Cape Riesling and occasionally South African Riesling and Paarl Riesling, and the variety is still increasingly popular with grape growers. It may be sold simply as Riesling within South Africa (where true Riesling is known as White or Weisser Riesling) and in some instances shares with that much greater German grape variety the ability to benefit from bottle ageing.

CRUCHINET, one of CHENIN BLANC's many synonyms.

🍇 **CURRANT,** variety used mainly for dried fruit but occasionally used for wine-making in Australia, where a variant on it, Carina, was bred in the 1960s.

D

🍇 **DATTIER,** Middle Eastern table grape grown in Australia as Waltham Cross and occasionally made into basic wine there.

🍇 **DEBINA,** sprightly, appley variety responsible for the lightly sparkling white wines of Zitsa in Epirus high in north west Greece near the Albanian border. It seems unlikely that the variety is not cultivated in Albania too. At these altitudes acidity levels remain high, and a tendency to oxidation has largely been checked by improved vinification methods.

🍇 **DELAWARE,** quantitatively important dark pink-skinned American hybrid that is probably derived from *Vitis labruscana, aestivalis,* and *vinifera*. It is popular in New York State and, for reasons that are now obscure, is the most widely planted in Japan. Its early ripening is presumably an advantage in Japan's damp autumns. The aromatic wine produced is not as markedly foxy as that of its great New York rival CONCORD, which ripens up to two weeks after it. It was first propagated in Delaware, Ohio in 1849.

🍇 **DIMIAT,** Bulgaria's most planted indigenous light-skinned variety, although it may not cover as much ground as RKATSITELI. It is grown mainly in the east and south of the country, where it is regarded as a producer of perfumed everyday whites of varying levels of sweetness but usefully dependable quality. The vines yield copper-coloured grapes in great quantity. The wines should be consumed young and cool. Dimiat is also a parent, with Riesling, of Bulgaria's MISKET.

🍇 **DINKA,** very ordinary but widely planted variety in Hungary and Vojvodina, where it covers a total of nearly 8,000 ha/20,000 acres. It is also known as Kövidinka, Kevedinka, and Ruzica.

🍷 **DOLCETTO,** early ripening, low-acid vine grown almost exclusively in the provinces of Cuneo and Alessandria in Piedmont, north west Italy. The wine produced is the local everyday lubrication, being gentle, fruity, and fragrant with flavours of liquorice and almonds. Most should be drunk within their first two or three years.

Because it can ripen up to four weeks before the majestic NEBBIOLO, Dolcetto can be grown on sites too high or too north-facing for Piedmont's greatest grape. In the revered Barolo and Barbaresco zones, for example, Dolcetto is rarely planted on a south-facing site unless the vineyard is too high to ripen Nebbiolo reliably. And in the zones of Dogliani, Diano d'Alba, and Ovada, Dolcetto is planted where other varieties may not ripen at all. There is a consensus amongst growers in the Dolcetto d'Alba zone, source of much of the finest Dolcetto, that the variety prefers the characteristic white marls of the right bank of the Tanaro and cannot give maximum results in heavier soils.

If the grape is relatively easy to cultivate, apart from its susceptibility to fungal diseases and a tendency to drop its bunches in the cold mornings of late September, it is far from easy to vinify. While low in acidity, relative to Barbera at least, and therefore 'dolce' (sweet) to the Piedmontese palate, Dolcetto (little sweet one) does have significant tannins which producers have tamed with shorter fermentations. So rich are the skins of Dolcetto in pigments (often resulting in heavy sediment) that even the shortest fermentation rarely compromises the deep ruby and purple tones of the wine.

There are seven Dolcetto DOCs in Piedmont: Acqui; Alba; Asti (where GRIGNOLINO is much more common); Diano d'Alba; Dogliani; the rare Langhe Monregalesi; and Ovada.

Ormeasco is Liguria's version of Dolcetto and is therefore the southernmost extent of Dolcetto territory in Italy. It grows just on the Ligurian side of the mountains that separate Piedmont from Liguria.

The French ampelographer Galet contends that Dolcetto is one and the same as the Douce Noire of Savoie, one of whose synonyms is Charbonneau, and that this is the variety known as CHARBONO in California. Not all Italians agree.

A variety known as Dolcetto is known in Argentina but it is grown on an extremely limited scale.

 DOMINA, modern German crossing that has had a certain success in Germany's cooler sites such as the Ahr and Franken, so that there were a total of 65 ha/160 acres planted by 1990. Its parents are Portugieser × Spätburgunder (Pinot Noir) and it combines the productivity of the first with the ripeness, tannins, and colour of the second but not its finesse and fruit.

 DOÑA BLANCA, also known as **Doña Branco** and Moza Fresca, Galician variety grown in north west Spain, particularly in Monterrei, Bierzo, and to a much lesser extent Valdeorras, where it is known as Valenciana. In the mid 1990s there were about 1,000 ha/2,500 acres and the slightly bitter wines of Monterrei suggest there may be potential.

 DORADILLO, fecund Australian immigrant, possibly from Spain. Its entirely unremarkable wine is more suitable for making brandy or basic fortified wines than for table wine. In the early 1990s there were still about 850 ha/2,100 acres of Doradillo planted, mainly in the hot, irrigated Riverland of South Australia. In its time it has erroneously been called Blanquette and considered identical to the JAÉN of Spain.

DORIN, rarely used name for CHASSELAS in the Vaud region of Switzerland.

 DORNFELDER, increasingly appreciated as Germany's most successful red wine crossing, bred mainly for its colour initially in 1956 by August Herold, who had unwisely already assigned his name to one of its parents, the lesser HEROLDREBE. Dornfelder therefore owes its name to the 19th century founder of the Württemberg viticultural school. A HELFENSTEINER × Heroldrebe cross, Dornfelder incorporates every important red wine vine grown in Germany somewhere in its genealogy and happily seems to have inherited many more of their good points than their bad.

The wine is notable for its depth of colour (useful in a country where pigments are at a premium), its good acidity but attractively aromatic fruit, and, in some cases, its ability to benefit from oak ageing and even to develop in bottle. It can often provide more drinking pleasure than a Spätburgunder (Pinot Noir), perhaps because its producers' ambitions are more limited. In the vineyard, the vine is easier to grow than Spätburgunder, has much better resistance to rot than Portugieser, stronger stalks than Trollinger, better ripeness levels than either, earlier ripening than Limberger (Blaufränkisch), and a yield that can easily reach 120 hl/ha (6.8 tons/acre) (although quality-conscious producers are careful to restrict productivity). It is hardly surprising that it continues to gain ground in most German wine regions, especially Rheinhessen, and Pfalz where results are particularly appetizing. Germany's total plantings rose steadily throughout the 1980s to reach a total of more than 1,200 ha/3,000 acres by the end of the decade.

The best bottles of this variety, usually sold as a varietal rather than blended, and either designed for youthful drinking or, in the case of oak-aged examples, for cellaring, demonstrate the point of vine breeding.

DOUCE NOIRE. See CHARBONO.

🍇 **DRUPEGGIO,** name for the light-berried CANAIOLO Bianco that adds interest to Trebbiano grapes, along with Grechetto, Malvasia, and Verdello, in the Orvieto wine of central Italy.

🍇 **DUNKELFELDER,** German TEINTURIER vine valued mainly for the depth of its colour, a useful commodity in Germany's blending vats. Plantings totalled rather more than 100 ha/250 acres in the early 1990s, mainly in the Pfalz and Baden.

🍇 **DURAS,** Gaillac speciality which gives deep-coloured wines with good structure but which has declined sharply in popularity because it buds dangerously early.

🍇 **DURELLA,** Veneto grape notable for its acidity and the basis of Lessini Durello.

🍇 **DURIF,** undistinguished variety now hardly grown in France, a selection of the PELOURSIN vine propagated eponymously by a Dr Durif in south eastern France in the 1880s. Valued for its resistance to downy mildew, not for the quality of its wine, it was tolerated but not encouraged by the French authorities in such regions as Isère and the Ardèche in the mid 20th century.

Today it has all but disappeared from France but was for long thought to be the same as the variety known in both North and South America as PETITE SIRAH. DNA fingerprinting at the University of California at Davis has since cast doubt on this hypothesis.

DUTCHESS

🍇 **DUTCHESS,** American hybrid based on *Vitis labrusca* and possibly *aestivalis* and *bourquiniana* grown with limited success in New York State. First identified in 1868 at Marlboro, New York.

E

🍇 **EARLY BURGUNDY,** California name for a variety once grown there in some quantity and identified by ampelographer Pierre Galet as ABOURIOU.

🍇 **EHRENFELSER,** one of the finest of German 20th century crossings, a Riesling × Silvaner developed at Geisenheim in 1929. In this case the aim of producing a super-Riesling that will ripen in a wider range of sites was achieved and the crossing's only inherent disadvantages are that the wine is slightly too low in acidity for long-term ageing—and that it cannot be called Riesling. Ehrenfelser, named after the Rheingau ruin of Schloss Ehrenfels, regularly ripens better and more productively than Riesling but is not nearly as versatile in terms of site as the more recently developed KERNER, which became a more obvious choice as a flexible Riesling substitute. Total plantings of Ehrenfelser fell in the late 1980s to less than 500 ha/1,200 acres, with the largest areas in the vine patchworks of Pfalz and Rheinhessen.

🍇 **ELBLING,** ancient, some would say outdated, vine that has been cultivated in the Mosel valley in northern Germany since Roman times. At one time it was effectively the only variety planted in Luxembourg, and dominated the extensive vineyards of medieval Germany. Today it is far less important than Rivaner (MÜLLER-THURGAU) in Luxembourg but retains a fairly constant 1,100 ha/2,700 acres of vineyard in Germany, most of it in the upper reaches of the Mosel-Saar-Ruwer above Trier, where chalk dominates slate and Riesling has difficulty ripening. Much of the Elbling grown here is used for sparkling Sekt and its naturally high acidity is certainly mitigated by the addition of carbon dioxide. While the vine is distinguished for its antiquity and productivity, its wines are distinguished by their often seering acidity and their relatively low alcohol, making it an even tarter, lighter version of SILVANER. (Weisser Silvaner is one of Elbling's German synonyms.) Must weights are typically only about 60 Oechsle, about 10 lower than Riesling. In the vineyard Elbling can produce up to 200hl/ha (11 tons/acre) but its productivity can vary as the vine is prone to coulure. Only the most dedicated wine-maker can extract any suggestion of Elbling's evanescent flavour of just-ripe apricots, but this is a wine to appeal to viticultural archivists.

Often grown as a vineyard mixture of Elbling Rouge and Elbling Blanc, the variety is called Räifrench in Luxembourg.

⚜ **EMERALD RIESLING,** one of the earliest of vine varieties developed at the University of California at Davis by Dr H. P. Olmo to emerge as a varietal wine. A Muscadelle × Riesling cross, it had its heyday in the late 1960s and early 1970s before slumping towards oblivion. Although just over 1,000 acres/400 ha remained in the early 1990s, mainly in the very south of the Central Valley, nearly all of the grapes disappear into generic blends. The purpose of the variety was to permit light, crisply acidic wines to be grown in warm inland climates. In spite of that aim, it performs best in the coastal counties, especially Monterey, however. Like its red stablemate RUBY CABERNET, it has also been tried with a certain degree of success in South Africa.

⚜ **EMPEROR,** hot climate table grape occasionally pressed into wine-making duty in Australia.

⚜ **ENCRUZADO,** Portuguese variety most commonly planted in Dão. It tends to yield small quantities of quite respectable perfumed wine.

⚜ **ERBALUCE,** speciality of Caluso in the north of the Piedmont region, north west Italy. Most dry Erbaluce is relatively light bodied and acidic, although there are some fine examples. Erbaluce's most famous, if rare, manifestation is the golden sweet apple-skin-flavoured Caluso Passito.

ERMITAGE, occasional synonym for MARSANNE.

⚜ **ESGANA CÃO,** occasionally just **Esgana,** synonyms for the Portuguese white grape variety known on the island of Madeira as SERCIAL. Its full name on the mainland means 'dog strangler', presumably a reference to its notably high acidity. It can be found as a particularly piercing ingredient in Vinho Verde, Bucelas, and white port.

⚜ **ESPADEIRO,** relative rarity grown in both the Rias Baixas zone of Galicia and across the border in Portugal's Vinho Verde country. It can produce quite heavily and rarely reaches high sugar levels.

The variety called Espadeiro near Lisbon is in fact TINTA AMARELA according to Mayson.

ESPARTE, old Australian name for MOURVÈDRE.

⚜ **ESQUITXAGOS,** common grape around Tarragona in eastern Spain which may be identical to MERSEGUERA.

⚜ **ÉTRAIRE DE L'ADUI, ÉTRAIRE DE LA DUI,** historic vine grown before phylloxera on the fringes of the south east Rhône valley and Savoie. Very similar to PERSAN.

⚜ **EZERJÓ,** Hungarian speciality in decline but still the country's fourth most important variety of either colour. Thin grape skins make them prey to rot. Most of the wine produced is relatively anodyne, but Móri Ezerjó produced from the vineyards near the town of Mór enjoys a certain following as a light, crisp, refreshing drink, and the Ezerjó grown, in quantity, in the far north west of Hungary can also yield lively dry whites for early consumption. Ezerjó means 'a thousand boons'.

F

🍇 **FABER, or FABERREBE,** like SCHEUREBE, a German crossing bred by Dr Scheu at Alzey in the Rheinhessen in the early 20th century. This crossing of Weissburgunder (Pinot Blanc) and Müller-Thurgau emerged in 1929. Faber, which will ripen easily in sites unsuitable for Riesling, has been particularly popular with growers in the Rheinhessen, where three-quarters of Germany's total of 2,000 ha/5,000 acres were grown in 1990, although total plantings are declining after a peak in the mid 1980s. Germans see Faber as a 'traditional' variety in that it shows some of the raciness of Riesling and has markedly more acidity than Müller-Thurgau. Acidity levels are even higher than those of Silvaner, which adds considerably to its appeal although the wines are not intensely flavoured, are not designed to age, and are generally most useful for blending. Faber can easily be persuaded to reach the must weights that qualify for Spätlese status if picked after Müller-Thurgau in a fine autumn. Like the even more popular BACCHUS, Faber suffers from stalk necrosis and, like the most popular modern crossing KERNER, needs careful trimming during the growing season. A little is grown in England.

🍇 **FALANGHINA, FALANGHINA GRECO,** very characterful ancient vine which may have provided a basis for the classical Falernum wine and is still grown on the coast of Campania north of Naples. Enjoying a small revival.

🍇 **FALSE PEDRO,** once used in South Africa for an Andalucian variety called Pedro Luis, and used in Australia for another southern Spanish vine called Cañocazo.

🍇 **FAVORITA,** speciality of Piedmont in north west Italy, often considered a relation of VERMENTINO because of the resemblance between the bunches of the two varieties. Favorita was not widely planted in the mid 1990s, having lost ground to ARNEIS in the Roero and to the newly popular CHARDONNAY in the Langhe. The wine itself has a pleasant citric tang, and its higher acidity means it can last longer than Arneis. The best examples have a distinctly alpine savour to them, together with considerable body—not unlike ROUSSANNE.

FENDANT, Valais name for the most planted grape variety in Switzerland, the productive CHASSELAS.

♀ FER, alias **Fer Servadou** (and many other aliases), now rare but old vine traditionally encouraged in many of the sturdy red wines of South West France. In Madiran, often called Pinenc, it is a distinctly minor ingredient, alongside Tannat and the two Cabernets. In Gaillac, called Brocol or Braucol, it has also lost ground. It is technically allowed into wines as far north as Bergerac, but today it is most important to the red wines of the Aveyron *département*, Entraygues, Estaing, and the defiantly smoky, rustic Marcillac. The iron hardness of the name refers to the vine's wood rather than the resulting wine, although it is well coloured, concentrated, and interestingly fumed with a hint of rhubarb. Fer has also been invited to join the already crowded party of varieties permitted in Cabardès.

The variety called Fer, of which there are about 1,500 ha/3,700 acres in Argentina, is apparently a clone of MALBEC.

♀ FERNÃO PIRES, versatile Portuguese variety that is the country's most planted, with more than 23,000 ha/57,000 acres in 1992. Its distinctive aroma can be somewhat reminiscent of boiled cabbage. A relatively early ripener and useful in blends, it is planted all over Portugal, notably in Ribatejo (where some oak-aged and botrytized sweet wines have been produced) and, as Maria Gomes, is the most common white grape variety in Bairrada, where it is often used for sparkling wines. The variety has also been planted experimentally in South Africa.

♀ FERRÓN, or **Ferrol,** minor but characterful Galician.

♀ FETEASCA, FETIASKA, or FETEASKA, scented white grape variety grown widely in eastern Europe, especially in Romania, where the two most planted varieties are the national speciality **Fetească Regală** followed by its parent **Fetească Albă**. Fetească Regală is Romania's own crossing of the GRASĂ of Cotnari and Fetească Albă developed in Daneş in Transylvania in the 1930s. Fetească Albă ripens earlier than the later crossing, which can be prone to rot. Both light-berried Feteascas are made into soft, peachy, aromatic, almost MUSCAT-like wines with definite if varying degrees of sweetness and, often, slightly too little acidity. More sophisticated vinification might well squeeze finer wines out of this variety eventually. Fetească Regală was Romania's most planted grape variety in 1993 with more than 17,000 ha/42,000 acres, when Fetească Albă was planted on more than 15,000 ha.

The vine is also important in Hungary, in Bulgaria, Moldova, and Ukraine. Fetească Regală is known as Királyleányka, while Fetească Albă is Leányka. When cited on a label for export to Germany, the grape's name is often directly translated as Mädchentraube, or Maiden's Grape. (In Romanian, Fetească meaning young girl's grape, contrasts directly with BĂBEASCĂ, grandmother's grape.)

There is also a dark-skinned variant, **Fetească Neagră**, whose red wines show potential when well vinified and carefully aged, but there were only about 1,000 ha/2,500 acres in Romania in the mid 1990s.

FEUILLE DE TILLEUL, synonym for Hungary's light-berried HÁRSLEVELŰ.

🍷**FIANO,** strongly flavoured classical vine responsible for Campania's Fiano di Avellino in southern Italy. Wines made from this variety are sturdy, are capable of developing for many years in bottle, and can mature from honeyed through spicy to nutty flavours (although old-fashioned wine-making has in some cases added its own notes of heaviness and premature oxidation).

🍷**FIÉ,** occasionally written **Fiét,** old Loire white grape variety thought to be an ancestor of SAUVIGNON BLANC. The variety has largely been abandoned because of its remarkably low yield, but producers such as Jacky Preys of Touraine pride themselves on their richer versions of this Sauvignon made from particularly old vines.

🍷**FINDLING,** mutation of MÜLLER-THURGAU grown to a limited extent in the Mosel-Saar-Ruwer of Germany, where its higher must weights are treasured, even if its tendency to rot is not.

🍇**FLAME SEEDLESS,** table grape variety occasionally used by wine-makers in California and Australia.

🍷**FLORA,** perhaps the most delicately aromatic of Dr H. P. Olmo's Davis creations (see also CARNELIAN, EMERALD RIESLING, RUBY CABERNET, SYMPHONY). Flora deserves a rather better fate than it had endured by the early 1990s when acreage was too small for official statistics and Flora rarely appeared as varietal wine. The easiest place to find it is in the sparkling Schramsberg Cremant. A result of Gewürztraminer × Sémillon, it apears to take after Gewürztraminer in cooler climates, Sémillon in warmer ones.

FOGONEU. See CALLET.

🍷**FOLGOSÃO,** early ripening, minor grape planted in the Douro and environs in northern Portugal.

🍷**FOLLE BLANCHE,** once grown in profusion along the Atlantic seaboard of western France, providing very acidic but otherwise neutral base wine for distillation. Phylloxera brought about its decline and France's total plantings of Folle Blanche continue to slide: from 12,000 ha/30,000 acres in 1968 to 3,500 ha 20 years later. It has also been grown to a very limited extent in California.

FOLLE NOIRE, is an occasional synonym for various French dark-berried varieties including JURANÇON and NÉGRETTE.

🍇 **FORASTERA,** common grape of the Spanish Canary Islands.

🍇 **FORCAYAT, FORCALLAT,** almost extinct grape of Valencia in south east Spain making curious-smelling but light-coloured wines for blending.

🍇 **FORTANA,** tart speciality of Emilia in north central Italy, known as Uva d'Oro in Romagna. Sometimes known as Fruttana, according to Anderson.

FRANCONIA, local name for BLAUFRÄNKISCH in Friuli, north east Italy.

FRANKEN RIESLING, occasional name for SILVANER.

FRANKOVKA, synonym for the red BLAUFRÄNKISCH grape used in Slovakia and Vojvodina.

FRANS, FRANSDRUIF, South African name for PALOMINO.

🍇 **FRAPPATO, FRAPPATO DI VITTORIA,** responsible for lightish fruity Sicilian reds, especially Cerasuolo di Vittoria, in the south east of the island.

🍇 **FREISA,** light red Piedmontese varietal, especially in Asti, Alessandria, and Cuneo. The vine was known in Piedmont in 1799 and at least two clones have been identified, the smaller-berried **Freisa Piccola** generally being planted on hillier sites and the larger-berried **Freisa Grossa** producing less lively wines on flatter vineyards. Freisa musts can be quite high in both acids and tannins, even if relatively light coloured for the region. Freisa is made in a range of styles, but the predominant, highly controversial one is slightly frothy, sweet *and* bitter with aromas of raspberries and violets. Very Piedmont. Some producers are experimenting with a more age-worthy, completely dry, and completely still type of Freisa. Italy's total plantings of Freisa fell by a third to 2,000 ha/5,000 acres in 1990.

Freisa is quite widely grown in Argentina, although total plantings are no more than a few hundred ha.

🍇 **FREISAMER,** 20th century German crossing (SILVANER × Ruländer or PINOT GRIS), originally called **Freiburger** after its birthplace. It reached a peak of popularity in the German wine region of Baden in the early 1970s but is declining in popularity, a casualty of KERNER's success perhaps. It is still grown today in north and central Switzerland, and sweeter versions are a speciality of the Herrschaft in Graubünden.

FRENCH COLOMBARD, common California name for COLOMBARD, now much more widely planted in California than in France.

FRONTIGNAC

FRONTIGNAC, FRONTIGNAN, synonyms for MUSCAT BLANC À PETITS GRAINS used particularly in Australia.

FRÜH is German for 'early'.

♥ FRÜHBURGUNDER, BLAUER, an early ripening strain of Spätburgunder (Pinot Noir) grown to a very limited extent in Württemberg. It tastes like a paler, leaner version of the lightest red burgundy.

♥ FRÜHROTER VELTLINER, or FRÜHER ROTER VELTLINER, early ripening red-skinned VELTLINER that most commonly makes white wines in Austria, where there were still about 800 ha/2,000 acres planted in the late 1980s, mainly in the Weinviertel region. The wine produced is often less distinguished than that made from Austria's most common grape variety GRÜNER VELTLINER, being notably lower in acidity in many cases. Yields are also lower.

In Germany in the early 1990s there were still a few ha of this variety, known as **Frühroter Malvasier** or occasionally Roter Malvasier, in the Rheinhessen region. Small plots may still be encountered in older vineyards of Alto Adige of Italy (where it is known as Veltliner) and Savoie (where it may be known as Malvoisie Rouge d'Italie).

FUMÉ BLANC, not a true vine variety synonym at all but a clever marketing spin on the Loire synonym Blanc Fumé for the SAUVIGNON BLANC.

♥ FUMIN, tough Valle d'Aosta speciality that is usually blended.

♥ FURMINT, fine, fiery grape variety grown most widely in the Tokaj region of north west Hungary and just over the border with Slovakia. It is also known, at least historically, in Austria's Burgenland, where it may be called Mosler. Relationships with many other varieties have been posited: Sipon in Slovenia, Poşip of Croatia, and GRASĂ of Romania.

Furmint is the principal ingredient in the great sweet Tokaji wines, the late ripening grapes being particularly sensitive to noble rot. Furmint wine is also characterized by very high acidity, which endows the wine with long ageing potential, high sugar levels and extract with particularly rich flavours. In Tokaj it is usually blended with up to half as much of the more aromatic grape variety HÁRSLEVELŰ. Some Muscat Lunel (MUSCAT BLANC À PETITS GRAINS) is also sometimes included in the blend.

Furmint also makes a characterful dry varietal to be drunk young and can easily produce wines with an alcoholic strength as high as 14 per cent. Total Hungarian plantings almost doubled between 1960 and 1990, when there were about 6,800 ha/16,800 acres, new plantings being restricted to approved clones T88, T92, Király Furmint, and Nemes Furmint. It is grown in Somló, Mescekalj, and Villány-Siklós but is most common in the Tokaj-Hegyalja region. The vine buds early but ripening slows towards the end of the season and the botrytized (*aszú*) grapes may not be picked until well into November in some years.

The vine is so well established in Hungary that there is little concrete evidence for its geographical origins.

The variety was also grown in Crimea when Tokay found such favour at the Imperial Court that the Tsars wished to make their own version, and it is still grown to a limited extent in South Africa, where it was imported in tandem with the other Tokay grape Hárslevelű.

G

🍇 **GAGLIOPPO,** predominant variety in Calabria in the far south of Italy for wines such as Cirò. Possibly of Greek origin, it thrives in dry conditions and reaches high sugar levels which result in robust, if rarely subtle, wines. It is also grown in Abruzzi, the Marches, and Umbria. Italy's total plantings of the variety were about 7,000 ha / 17,000 acres in 1990. According to Burton Anderson it is called variously Arvino, Lacrima Nera, Magliocco, Mantonico Nero, and even sometimes AGLIANICO, to which it may be related.

🍇 **GALEGO DOURADO,** grape grown on the Atlantic coast of Portugal and known for its high alcohol wines.

🍇 **GAMASHARA,** Azerbaijan speciality, along with MATRASSA.

🍇 **GAMAY,** the Beaujolais grape. Galet cites scores of different Gamays, many quite unrelated to the Beaujolais archetype, many of them particular clonal selections of it, and many more of them red-fleshed TEINTURIERS once widely used to add colour to vapid blends. Even as recently as the 1980s, more than 1,000 ha / 2,500 acres of Gamay Teinturiers were planted in France, and they can still be found in the Mâconnais and the Touraine. The 'real' Gamay is officially known as **Gamay Noir à Jus Blanc** to draw attention to its noble white flesh.

The vine is a hasty one, budding, flowering, and ripening early, which makes it prone to spring frosts but means that it can flourish in regions as cool as much of the Loire. It can easily produce too generously and the traditional gobelet method of training (without wires or stakes) is designed to match this aptitude to the granitic soils of the better Beaujolais vineyards.

Gamay juice also tends to be vinified in a hurry, not least because of market pressure for Beaujolais Nouveau, and if Gamay-based wines are cellared for more than two or three years it is usually by mistake. As a wine Gamay tends to be paler and bluer than most other reds, with relatively high acidity and a simple but vivacious aroma of freshly picked red fruits, often overlaid by the less subtle smells associated with rapid, oxygen-free fermentation such as bananas, boiled sweets, and acetate. In France and Switzerland it is often blended with Pinot Noir, endowing the nobler grape with some precocity, but often blurring the very distinct attributes of each. Gamay fruit is naturally low in potential alcohol, and for many Gamay's charm lies in its refreshing lightness, but prevailing perceptions equating weight with worth encouraged many wine-makers to add sugar routinely at fermentation to make more alcoholic wines.

Gamay and Beaujolais are entirely interdependent. No wine region is so determinedly single-minded as Beaujolais, which has just a few Chardonnay vines for Beaujolais Blanc to prove the rule, while 33,600 ha / 83,000 acres in 1988, well over half of the world's total Gamay plantings, are in this single region. Similar, often lighter and arguably truer, wines are made from the Gamay grown in the small wine regions of central France, particularly those around Lyons and in the upper reaches of the Loire such as Châteaumeillant, Coteaux du Lyonnais, Coteaux du Giennois, Côtes d'Auvergne, Côtes du Forez, Côtes Roannaises, and St-Pourçain.

Outside Beaujolais, and perhaps because its wines have been seen as too different from the intense, tannic, fashionable norm, the Gamay vine has been losing ground. In the Côte Chalonnaise and the Mâconnais between Beaujolais and the Côte d'Or Gamay was displaced as principal grape variety by Chardonnay during the 1980s, and the unexcitingly muddy quality of Gamays made here is expected to continue this trend. Gamay still took up 400 ha / 1,000 acres of the Côte d'Or's valuable vineyard in 1988 but is fast being supplanted by more rewarding varieties.

Gamay is grown all over the Loire but is not glorified by any of the Loire's greatest appellations. Gamay de Touraine can provide a light, sometimes acid but usually cheaper alternative to Beaujolais, but it is most widely grown west of the Touraine, alongside Sauvignon, for such light, lesser-known names as Cheverny and Coteaux du Vendômois. Gamay also provides about 40 per cent of all of the Loire's important generic Vin de Pays du Jardin de la France.

It is also grown in minute quantities in Canada and is confused on a grand scale with BLAUFRÄNKISCH throughout eastern Europe. It is grown to a certain extent in Italy, and plays a relatively important role in the vineyards of what was Yugoslavia, notably in Croatia, Serbia, Kosovo, and, to least effect, in Macedonia.

It is chiefly valued, however, outside Beaujolais, by the Swiss who grow it widely, most enthusiastically around Geneva. It thrives well at high altitudes and is commonly blended with Pinot Noir to become Dôle in Valais and Salvagnin in Vaud. Too often, however, the Swiss are apt to chaptalize the life out of it.

Outside France there is even less incentive to develop this under-appreciated variety although a few California growers have bothered to import and vinify true Gamay as opposed to the less distinguished vine known there as **Napa Gamay** (see VALDIGUIÉ) or the variety called **Gamay Beaujolais**, which is in fact a lesser clone of Pinot Noir.

GAMÉ, BLAUFRÄNKISCH in Bulgaria.

GAMZA, Bulgarian name for Hungary's KADARKA.

🍇**GARGANEGA**, vigorous, productive, often over-productive vine grown in the Veneto region of north east Italy, most famously for Soave, in which it may constitute anything from 70 to 100 per cent of the blend, often sharpened up by the addition of TREBBIANO di Soave, but increasingly plumped up by CHARDONNAY and other imports. In the original heart of the zone, with yields kept well in check, it can produce the fine, delicate whites redolent of lemon and almonds which give Soave a good name. The vine is also responsible for Gambellara—indeed **Garganega di Gambellara** is its most important subvariety. Garganega has such a long history in the Veneto that it has developed myriad, if rarely particularly interesting, strains, clones, and subvarieties. Other wines in which it plays a major part include Bianco di Custoza, Colli Berici, Colli Euganei, and it is also grown to a more limited extent in both Friuli and Umbria. It is Italy's fifth most important white grape variety, planted on more than 13,000 ha/32,000 acres in 1990.

🍇**GARNACHA**, the Spanish, and therefore original, name for the grape known in France and elsewhere as GRENACHE. Its most common and noblest form is the black-berried and white-fleshed **Garnacha Tinta**, sometimes known as **Garnacho Tinto**.

In Spain Garnacha Tinta is grown extensively, particularly in north and east, being an important variety in such wine regions as Rioja, Navarre, Ampurdàn-Costa-Brava, Campo de Borja, Cariñena, Costers del Segre, Madrid, La Mancha, Méntrida, Penedès, Priorato, Somontano, Tarragona, Terra Alta, Utiel-Requena, and Valdeorras. In Rioja it provides stuffing and immediate charm when blended with the more austere TEMPRANILLO. The cooler, higher vineyards of Rioja Alta are reserved for Tempranillo, while Garnacha is the most common grape variety of the warm eastern Rioja Baja region where the vines can enjoy a long ripening season. The juiciness apparent in these early maturing riojas can be tasted in a host of other Spanish reds and, especially, rosados. Grenache has been adopted with particular enthusiasm in Navarre, where it has been the dominant grape variety and dictates a lighter, more obviously fruity style of red and rosado than in Rioja. The authorities, anxious to modernize Navarre's image, have been positively discouraging new plantings of Garnacha, however.

Perhaps the most distinctive Spanish wine based on Garnacha Tinta (often incorporating some **Garnacha Peluda,** or 'downy Garnacha', known as LLADONER PELUT in Languedoc-Roussillon), is Priorato, the concentrated Catalonian cult wine in which the produce of old Garnacha vines may be modernized by blending it with young Merlot, Cabernet, or even Syrah fruit. ▭▬▭

Garnacha Tintorera, on the other hand, is a synonym for the red-fleshed ALICANTE BOUSCHET, Tintorera being Spanish for 'dyer' or TEINTURIER.

Garnacha Blanca is the light-berried Grenache Blanc (see GRENACHE).

🍷 GARRIDO, minor speciality of the Condado de Huelva region in southern Spain.

🍷 GARRUT, rare Catalonian variety producing aromatic wines reminiscent of liquorice that are high in tannins. ▭▬▭

🍷 GEWÜRZTRAMINER, often written **Gewurztraminer,** pink-skinned grape variety responsible for particularly pungent, full bodied white wines. Gewürztraminer may not be easy to spell, even for wine merchants, but is blissfully easy to recognize—indeed many wine drinkers find it is the first, possibly only, grape variety they are able to recognize from the wine's heady perfume alone. Deeply coloured, opulently aromatic, and fuller bodied than almost any other white wine, Gewürztraminer's faults are only in having too much of everything. It is easy to tire of its weight and its exotic flavour of lychees and heavily scented roses, although Alsace's finest Gewürztraminers are extremely serious wines, capable of at least medium-term ageing.

This by now internationally famous vine variety's genealogy is both ramified and fascinating. Traminer, like Gewürztraminer but with pale green berries and much less scent, is the original variety, first noted in the village of Tramin or Termeno in what is now the Italian Tyrol around 1000 AD. It was popular here until the 16th century when the much more ordinary, but more prolific, VERNATSCH, or Schiava, supplanted it.

Traminer has also been known in Alsace since the Middle Ages, although it is said to be cuttings imported much more recently from the Pfalz, the German wine region in which it was widely grown and prized for its richness, that encouraged its spread in Alsace. Today, according to Galet, the SAVAGNIN so vital to the *vin jaune* of the Jura is none other than Traminer, both varieties famous for their ripeness levels, depth of flavour, and ability to age.

Traminer, like PINOT, mutates easily, however, and Gewürztraminer is the name adopted in the late 19th century for the dark pink-berried MUSQUÉ mutation of Traminer (and adopted as its official name in Alsace in 1973). Although much has been read into the direct German translation of *gewürz* as 'spiced', in this context it simply means 'perfumed'. Traminer Musqué, Traminer Parfumé, and Traminer Aromatique were all at one time French synonyms for Gewürztraminer. As early as 1909 the ampelographer Viala acknowledged Gewürztraminer as an accepted synonym for Savagnin Rosé, and this aromatic, dark-berried version is known as Roter Traminer in German and

Traminer or Termeno Aromatico, Traminer Rosé, or Rosso in Italy. Its long history in Alsace means that it is occasionally known as some sort of CLEVNER, particularly in this case Rotclevner.

Gewürztraminer has become by far the most planted variant of Traminer. The grapes are certainly notable at harvest for their variegated but incontrovertibly pink colour, which is translated into very deep golden wines, sometimes with a slight coppery tinge. Wine-makers unfamiliar with the variety have been known to be panicked into subtracting colour, and flavour. Gewürztraminers also attain higher alcohol levels than most white wines, with over 13 per cent being by no means uncommon, and acidities can correspondingly be precariously low. The second, softening malolactic fermentation is almost invariably suppressed for Gewürztraminer and steps must be taken to avoid oxidation.

If all goes well the result is deep golden, full bodied wines with a substantial spine and concentrated heady aromas (almost reminiscent of bacon fat in dry versions) whose acidity level will preserve it while those aromas unfurl. In a lesser year or too hot a climate the result is either an early picked, neutral wine or an oppressively oily, flabby one that can easily taste bitter to boot.

Viticulturally, Gewürztraminer is not exactly a dream to grow. Relative to the varieties with which it is commonly planted, it has small bunches and is not particularly productive, although the Germans have predictably selected some high-yielding clones. Its early budding leaves it prey to spring frosts and it is particularly prone to virus diseases, although the viticultural station at Colmar has developed such virus-free clones as those numbered 47, 48, and 643.

Since Gewürztraminer has been seen as a second rank variety in terms of international popularity and saleability (perhaps because of its associations with the unfashionable Riesling), few wine-makers outside Alsace have expended real energy on making great Gewurz. By the early 1990s the finest examples still came almost exclusively from this region in eastern France where Gewürztraminer, Riesling, and PINOT GRIS are considered the only 'noble' grape varieties.

Of these three, Gewürztraminer was the second most planted in Alsace as a whole, and the most widely planted in the more famous vineyards of the Haut-Rhin *département*, with a total of 2,500 ha/6,100 acres in 1989. It is particularly successful on the richer clay soils of the Haut-Rhin and has inspired a raft of late harvest examples labelled Vendange Tardive or even Sélection de Grains Nobles in the sunnier harvests of the 1980s and 1990s. The variety easily attains must weights well in excess of Riesling at comparable ripeness levels and regulations take account of this. Such late harvest Gewürztraminers may not last the same number of decades as their Riesling counterparts, but many last longer than their first decade.

Earlier-picked Alsace Gewürztraminer should be intriguingly aromatic yet dry and sturdy enough to accompany savoury food but too many examples are simply scented fly-by-nights, lightweight wines produced from heavily cropped vines that taste as though they have been aromatized by a drop of MUSCAT OTTONEL. In these lower ranks, it can be difficult to distinguish a poor Alsace wine labelled Gewürztraminer from one labelled Muscat.

Germany relegates its (Roter) Traminer to a very minor rank, well behind Riesling, with much less than 1,000 ha in total, including some plantings of the non-aromatic sort, which is very occasionally bottled separately. The variety needs relatively warm sites to avoid spring frost damage and to assure good fruit set so that in northern Germany Riesling is usually a more profitable choice. More than half of Germany's Traminer is planted in Baden and Pfalz, where it can produce wines of discernible character but is too often associated with somewhat oily sickliness.

There is almost as much Traminer planted in Austria as in Germany but here too it has been consigned to the non-modish wilderness, even though some examples, particularly later-picked sweet wines from Styria, can exhibit an exciting blend of race and aroma and can develop for a few years in bottle.

The variety is grown, in no great quantity but usually distinctively, throughout eastern Europe, called Tramini in Hungary; Traminac in Slovenia; Drumin, Pinat Cervena, or Liwora in what was Czechoslovakia; occasionally Rusa in Romania and Mala Dinka in Bulgaria. Most of the vines are the aromatic mutation and demonstrate some of Gewürztraminer's distinctive perfume but often in extremely dilute, and often sullied, form, typically overlaying a relatively sweet, lightish white. Hungarians are particularly proud of their Tramini grown on the rich shores of Lake Balaton. It is grown by the Romanians in Transylvania, by the Bulgarians in the south and east and also, as Traminer, in Russia, Moldova, and Ukraine, where it is sometimes used to perfume sparkling wine.

It is grown in small quantities, sometimes called Heida, Heiden, or Païen in Switzerland, and in ever smaller quantity in Luxembourg. In Iberia Torres grow it in the High Penedès for their Viña Esmeralda and it is essentially a mountain grape even in Italy, where Traminer Aromatico is grown almost exclusively, and decreasingly, in its seat, Alto Adige. The less scented and less interesting Traminer is also grown to a limited extent, and Italian wine-making together with vineyard altitude do nothing to emphasize Gewürztraminer characteristics in the resulting wines, although the international nature of the variety may encourage a small renaissance of popularity.

In the New World Gewürztraminer presents a challenge. Many wine regions are simply too warm to produce wine with sufficient acidity, unless the grapes are picked so early, as in some of Australia's irrigated vineyards, that they have developed little Gewürztraminer character. Australia's 'Traminer' vine population, concentrated in some of the less exciting corners of South Australia and New South Wales, had fallen to 600 ha/1,500 acres by the early 1990s, much of it used to perfume and sweeten Riesling in commercial blends.

The variety has been more obviously successful in the cooler climate of New Zealand, although even here total plantings are falling, to 200 ha in 1990, the most lively wines coming from Gisborne on the east coast of the North Island. This, curiously, was one of the earlier identifications of varietal/geographical matching in the southern hemisphere.

Another happy home for Gewürztraminer is in the Pacific Northwest of America, particularly in Washington and Oregon, although even here the variety has been losing ground to the inevitable Chardonnay. Washington had 314 acres/127 ha in 1994 but could demonstrate some appetizing life in several well-vinified examples, even if too many were too sweet. In Oregon too the smoky fume of Alsace is apparent in some bottlings, although rot can be a problem in this wetter climate.

Gewürztraminer remains a relatively minor variety in California, however, whose 1,700 acres/700 ha, almost half of them in Monterey, too often bring forth oil rather than aroma. There are a few ha of Traminer in Argentina but generally South America relies on TORRONTÉS and MOSCATEL to provide aromatic whites. Limited plantings in South Africa have so far yielded sweet wines but only some of the right aromas.

It seems likely that serious Gewürztraminer will remain an Alsace speciality for some years yet.

☙ **GIRÒ,** Sardinian grape making some fortified wines. Probably brought from Spain in the early 15th century, according to Anderson.

☙ **GM 6494,** red-fleshed hybrid with Mongolian *Vitis amurensis* genes designed to withstand cold winters such as those experienced in England. It has small berries and can make light, fruity wines with a good ruby colour.

☙ **GODELLO,** fine Galician variety most successful in Valdeorras in north west Spain, where plantings are increasing once more. The wines are high in acidity but have a refreshing green apple flavour.

☙ **GOLDBURGER,** Austrian gold-skinned grape variety, a crossing of WELSCHRIESLING and Orangetraube. In the late 1980s there were 500 ha/1,200 acres of this vine, almost all of them in Burgenland where it ripens impressively but rarely produces exciting wines. Austria's answer to the many German crossings.

GOLDEN CHASSELAS, occasional synonym for CHASSELAS when reared as a table grape, and once the name used for California's PALOMINO plantings.

GOLDMUSKATELLER, name for golden-berried form of MUSCAT BLANC À PETITS GRAINS in the Italian Tyrol.

GORDO

GORDO, GORDO BLANCO, originally Spanish synonyms for MUSCAT OF ALEXANDRIA more recently adopted by Australians.

GOUVEIO, Douro white grape variety thought locally to be the same as the VERDELHO of Madeira. It ripens well but does not produce particularly distinguished wines in the Douro.

🌱 **GRACIANO,** sometimes called **Graciana,** is a richly coloured, perfumed variety once widely grown in Rioja in northern Spain. It has fallen from favour because of its inconveniently low yields, thereby depriving modern Rioja of an important flavour ingredient. It is still planted in Rioja (on about half a per cent of available vineyard) and is being encouraged in Navarre.

The vine buds very late and is prone to downy mildew but can produce wine of great character and extract, albeit quite tannic in youth. Known as Morrastel in France, it was popular in the Midi until the middle of the 19th century when Henri BOUSCHET stepped in to provide growers with a more productive, more disease-resistant, but wildly inferior crossing of Morrastel with Petit Bouschet, the Morrastel-Bouschet which in time virtually replaced all of the original Morrastel in French vineyards. True Morrastel, or Graciano, is still grown in southern France in minute quantities but Languedoc's viticultural archivists have recently shown interest in it. Morrastel-Bouschet, on the other hand, still covered 1,600 ha/4,000 acres of the Midi in 1979 but is rapidly being pulled up.

There is ample possibility for confusion since the Spaniards use the name Morrastel as a synonym for their very different, and widely planted, variety Monastrell (MOURVÈDRE). Even today in North Africa the name Morrastel is used for both Graciano and Mourvèdre.

The variety known as Xeres in California, which has also been planted on a similarly limited scale in Australia, is probably Graciano, as is Graciana, the variety of which there were 144 ha/356 acres recorded in Mendoza in 1989. Argentina therefore has the distinction of being home to the world's largest plantation of this interesting grape variety.

🌱 **GRAISSE,** also known as **Plant de Graisse,** minor variety grown in armagnac country where, unusually, it is capable of producing wine quite drinkable in pre-distilled form. The grape pulp is unusually viscous, however, and only a small area of this variety remains.

GRANACCIA, GRANACHA, Italian names for local strains of GRENACHE.

GRANDE VIDURE, historic synonym recently adopted by the Carmen winery of Chile for its plantings of CARMENÈRE identified in the early 1990s.

🍇 **GRAND NOIR DE LA CALMETTE**, hardly deserves its name, which suggests it is the great black grape variety of the BOUSCHET experimental vine-breeding station, Domaine de la Calmette. At least ALICANTE BOUSCHET had the relatively noble Grenache as a parent while 'Grand' Noir was bred from Petit Bouschet and the common ARAMON. Not surprisingly it has a very high yield and, from its TEINTURIER parent, red flesh (although not as red as Alicante Bouschet's).

Often known simply as **Grand Noir**, it was widely planted in France until the 1920s, when its susceptibility to powdery mildew and winter cold precipitated a decline in its fortunes. Although 2,000 ha / 5,000 acres remained in France at the beginning of the 1980s, the variety is rapidly being pulled up. It is still grown to a very limited extent in north east Victoria in Australia. See also Spain's GRAN NEGRO.

🍇 **GRAN NEGRO**, **GRÃO NEGRO**, rustic, red-fleshed grape grown to a limited extent in Valdeorras, north west Spain. Probably identical to France's GRAND NOIR for it makes deep-coloured wine and was introduced to Galicia after phylloxera.

GRAPPUT, synonym for the rapidly declining French grape variety BOUCHALÈS.

🍇 **GRASĂ**, the 'fat' white grape of Cotnari in Romania where it is grown exclusively, on a total of about 850 ha / 2,000 acres of vineyard in the early 1990s. It can reach extremely high must weights but needs the balancing acidity of grapes such as TĂMÎIOASĂ ROMÂNEASCĂ in a blend. In 1958 Grasă grapes in Cotnari reached a sugar concentration of 520 g/l. The vine is usefully sensitive to noble rot and is said to have been grown in this part of western Moldavia since the 15th century. Grasă is also a parent of FETEASCĂ Regală, Romania's most planted vine.

GRAȘEVINA, Croatian name for WELSCHRIESLING, the single most planted vine in former Yugoslavia. **Grassica** may also be encountered.

GRAU, German for grey or *gris*.

GRAUBURGUNDER, German synonym for PINOT GRIS used by many, particularly in Baden and Pfalz regions, to designate a crisp, dry style of wine as opposed to sweeter wines normally labelled with the variety's more common German synonym RULÄNDER.

🍇 **GRAY RIESLING**, California name for a pale grape variety that has nothing to do with Riesling and has been identified as a light-berried mutation of TROUSSEAU, Trousseau Gris. It makes pale-hued, pale-flavoured varietal white wines of pleasant but resolutely small character. Such wines, usually made off dry, were once popular but the variety covered fewer than 300 acres / 120 ha of California in the early 1990s.

GRECANICO DORATO

🍇 **GRECANICO DORATO,** Sicilian variety whose total vineyard area increased from less than 3,000 ha/7,500 acres to more than 4,500 ha during the 1980s. The name suggests Greek origins and the current wines may not be maximizing full aromatic, rather SAUVIGNON-like potential.

🍇 **GRECHETTO,** sometimes **Greghetto,** characterful, often astringent, central Italian variety most closely associated with Umbria, where it is a firm, full bodied, nutty, sometimes almost varnish-scented, ingredient in Orvieto and in the whites of Torgiano. The vine has good resistance to downy mildew and is sufficiently sturdy to make good vin santo. It is typically blended with TREBBIANO, VERDELLO, and MALVASIA, although in Antinori's most admired white wine Cervaro it has played a supporting role to CHARDONNAY. Occasionally called **Greco Spoletino** or **Greco Bianco di Perugia,** it is by no means identical to GRECO (although it may well share its Greek origins).

🍇 **GRECO,** name of one or perhaps several, usually noble, white grape varieties of Greek origin currently grown in southern Italy. According to the 1990 vineyard survey of Italy, **Greco Bianco** plantings totalled less than 1,000 ha/2,500 acres.

In Campania it produces the respected full bodied dry white **Greco di Tufo** around the village of Tufo, while, blended with FALANGHINA and BIANCOLELLA grapes, it makes a contribution to the inconsequential dry whites of the island of Capri.

Perhaps the finest Greco-based wine is the sweet **Greco di Bianco** made from semi-dried grapes grown around the town of Bianco on the south coast of Calabria.

Greco in various forms is also used as a synonym for the unrelated ALBANA.

🍇 **GRECO NERO,** vine variety, only possibly of Greek origin, of which there were 3,200 ha/7,900 acres planted in southern Italy in the early 1990s, mainly in Calabria, where it is often blended with GAGLIOPPO. Its local synonym is Marsigliana, according to Burton Anderson.

🍇 **GREEN HUNGARIAN,** undistinguished California speciality making bland wine in decreasing quantities. Galet claims that the variety is indeed Hungarian in origin and that it has been known and grown in both Alsace and the Gers in South West France as Putzscheere, a German name that refers to the beauty and abundance of its grapes (which have good rot resistance, according to Galet).

🍇 **GRENACHE,** the world's second most widely planted grape variety sprawling, in several hues, all over Spain and southern France. It probably owes its early dispersal around the western Mediterranean to the strength and extent of the Aragon kingdom, but it would make a rewarding subject for an ampelographical sleuth. It has been widely accepted that, as Garnacha, it originated in Spain in the northern province of Aragon and then spread to Rioja and Navarre before colonizing extensive vineyard land both north and south of the Pyrenees, notably in Roussillon, which was ruled by Spain, and more particularly by the kingdom of Aragon, for four centuries until 1659. From here Grenache made its way east. It reached Languedoc in the early 18th century and the southern Rhône by the 19th century. Grenache is undoubtedly, however, the same as Sardinia's CANNONAU, which the Sardinians claim as their own, advancing the theory that the variety made its way from this island off Italy to Spain when Sardinia was under Aragon rule, from 1297 until 1713.

Whatever its origins, Grenache covers more vine-dedicated ground than any grape variety other than AIRÉN, most commonly encountered in its darkest-berried form as GARNACHA Tinta, under which more details of its performance in Spain may be found. It is Spain's most planted black grape variety with more than 100,000 ha / 250,000 acres and the French census of 1988 demonstrated **Grenache Noir**'s continued advance on the Midi with a total of 87,000 ha by then. For a variety that covers so much terrain, it is remarkably rarely encountered by name by the wine drinker, much of it being blended with other varieties higher in colour and tannin.

With its strong wood and upright growth, Grenache Noir is well suited to traditional bushlike viticulture in hot, dry, windy vineyards. It buds early and, in regions allowing a relatively long growing cycle, can achieve heady sugar levels. The wine produced is, typically, paler than most reds (although low yields tend to concentrate the pigments in Spain), with a tendency to oxidize early, a certain rusticity, and more than a hint of sweetness. If the vine is irrigated, as it has tended to be in the New World, it may lose even these taste characteristics. If however, as by the most punctilious Châteauneuf-du-Pape producers, it is pruned severely on the poorest of soils and allowed to reach full maturity of both vine and grape, it can produce excitingly dense meaty, spicy reds that demand several decades' cellaring. The rediscovery of Rhône reds in the late 1980s encouraged some New World producers to invest more effort in their own Grenache, even though its sturdy trunk has made it less widely popular in the modern era of mechanical harvesting.

In France, the great majority of Grenache's extensive vineyards are in the windswept southern Rhône, where seas of Côtes-du-Rhône of varying degrees of distinction are produced alongside smaller quantities of Châteauneuf-du-Pape, Gigondas, Vacqueyras, and the like. It is undoubtedly the Grenache ingredient that determined Châteauneuf-du-Pape's unusual official requirement of a minimum alcoholic degree (of 12.5°). Although blending has been the watchword here, notably with the more structured Syrah, such exceptions as the famously concentrated Châteauneuf-du-Pape Ch Rayas show what can be done by low yields, determination, and Grenache alone. Grenache is also responsible for much of southern France's rosé, most obviously and traditionally in Tavel and in neighbouring Lirac but also much further eastwards into Provence proper. In the Languedoc Grenache plays an unsung supporting role in the blends of its appellation wines, but in Roussillon it is extremely important as the vital ingredient in such distinctive strong, sweet wines as Banyuls, Rivesaltes, and Maury—further proof that Grenache is capable of producing great wine, albeit a very particular sort of wine.

Grenache Noir is being uprooted in Corsica but in Sardinia, as Cannonau, it plays a dominant role in the island's reds, which can achieve daunting levels of natural ripeness, whether in deep, dark dry wines that may have as many as 15 degrees of natural alcohol or dessert wines. The vine is also grown in Calabria and Sicily.

Grenache's ability to withstand drought and heat made it a popular choice with New World growers when fashion had little effect on market forces. It can be difficult to ripen in coastal counties but thanks to extensive historic acreage in the Central Valley, and some in Mendocino, it was still California's third most planted black grape variety, with nearly 13,000 acres/5,300 ha, after Zinfandel and Cabernet Sauvignon in 1991, but was overtaken by fashionable Merlot soon after. Not even California's 'Rhône Rangers' are expected to rescue Grenache from this downward trend, for most vines ripen unevenly and produce thin, jammy stuff although dry-farmed, short-pruned fruit from old vines has been sought out by some. Its fortunes suffered a brief reprise in the late 1980s when White Grenache (made to 'blush' from black grapes) was developed as the natural alternative to White Zinfandel when cheap ZINFANDEL grapes were in short supply.

Grenache was Australia's most planted black grape variety until the mid 1960s. SHIRAZ (Syrah) overtook it in the late 1970s but it was not until the early 1990s that Australia's Cabernet Sauvignon output overtook that of Grenache. The variety has been shamelessly degraded and milked in the heavily irrigated, undistinguished vineyards in which it has been expected to produce large quantities of wine for basic blends. An increasing band of producers, mainly in the Barossa Valley, now take it seriously, however, seeking out old bush vine plantings for special, concentrated bottlings, often blended with Shiraz and occasionally MOURVÈDRE to emulate Châteauneuf-du-Pape.

Grenache Noir is also grown in Israel, where it was exported at the end of the 19th century, to a very limited degree in South Africa and still, to a much greater extent, in North Africa where it was once an important element in the usefully soupy reds of Algeria and in some fine Moroccan rosés.

🍇 **GRENACHE BLANC**, the white-berried form of GRENACHE Noir, is discreetly important in southern France, where it was overtaken by Sauvignon Blanc as fourth most planted white grape variety only in the late 1980s. Although in decline, the variety is much planted in Roussillon, where it produces fat, soft white table wines. It is also an important ingredient in the paler Rivesaltes. Grenache Blanc is often encountered in the blended white wines of the Languedoc, to which it can add supple fruit if not longevity. It need not necessarily be consigned to the blending vat, however, and, since the early 1990s, flattering, soft, supple, almost blowsy varietal versions have been marketed. If carefully pruned and vinified, it can produce richly flavoured, full bodied wines that share some characteristics with MARSANNE and can even be worthy of ageing in small oak barrels.

As **Garnacha Blanca,** it plays a role in north eastern Spanish whites such as those of Alella, Priorato, Tarragona, Rioja, and Navarre.

Grenache Rose and **Grenache Gris** are also commonly encountered in southern French whites and some pale rosés.

♈ GRIGNOLINO, very localized Piedmontese grape variety of north west Italy sold almost invariably as a pale red varietal wine with an almost alpine scent and tang. Grignolino is a native of the Monferrato hills between Asti and Casale and serves the same function as DOLCETTO in the province of Cuneo: that of providing a wine that can be drunk young with pleasure while the more serious wines of the zone mature—although Grignolino is more difficult to match with food than the fuller Dolcetto. The light colour and relatively low alcohol (11 to 12 per cent) can be deceptive; the wine draws significant tannins from the abundant pips of the Grignolino grape. It takes its name from *grignole*, the dialect name for pips in the province of Asti.

Although Piedmont's producers have been regularly predicting a breakthrough for Grignolino that would transform it into Italy's answer to Beaujolais, the wine remains an unquestionably local taste and, with its rather odd combination of pale colour and perceptible acidity and tannins, somewhat *sui generis*. Grignolino seems to show a decided preference for dry, loose soils but is more frequently planted in heavier, moister ones.

The future of the vine appears uncertain, partly because of its extreme susceptibility to disease and, perhaps even more importantly, its tendency to ripen late and unevenly. This means that it requires the best sites and exposures which, in its home base of Monferrato, are increasingly being reserved for BARBERA or, in the case of the highest vineyards, for international white varieties such as CHARDONNAY. The 1990 survey found only 1,350 ha/3,300 acres of Grignolino, although this represented a small increase on the 1982 area.

A variety called Grignolino is also grown to a very limited extent in California, where its rosé has a certain following and a port-like version is not unknown.

♈ GRILLO, Sicilian variety once the base of Marsala. Grillo may have potential for table wines but plantings halved in the 1980s so that there were hardly more than 2,000 ha/5,000 acres in 1990.

Citrus-flavoured, full bodied wines can demonstrate a certain earthiness, even astringency, and can respond well to barrel ageing.

GRINGET. See SAVAGNIN.

🍇 **GROLLEAU,** the Loire's everyday dark-berried vine, producing extremely high yields of relatively thin, acid wine. Fortunately, it is being systematically replaced with GAMAY and, more recently, CABERNET FRANC. Total French plantings are falling steadily but were still nearly 4,000 ha/10,000 acres in 1988. The status of the variety is such that it is allowed into the rosé but not red versions of appellation contrôlée wines such as Anjou, Saumur, and Touraine. It has played a major part only in Rosé d'Anjou, in which it is commonly blended with Gamay (which ripens just before it) and is a valued ingredient in some sparkling wines.

Grolleau Gris produces innocuous white wines which may form part of the blend for the Loire's regional Vin de Pays du Jardin de la France, although if yields are restricted (which can be very difficult) it can produce varietal wines of real character.

🍇 **GROPPELLO,** variety grown to a limited extent on the shores of Lake Garda in Italy producing lightish reds.

GROSLOT, alternative spelling of the Loire's GROLLEAU.

🍇 **GROS MANSENG,** Basque variety grown in South West France to produce mainly drier versions of Jurançon and various Béarn wines. It is also now allowed into Gascony's Pacherenc du Vic Bilh. The vine looks similar to but is distinct from PETIT MANSENG. It yields more generously and produces discernibly less elegant, less rich, but still powerful wine. Unlike the smaller-berried Petit Manseng, it is not sensitive to coulure. See MANSENG for some details of the area planted with both varieties.

GROS RHIN, Swiss synonym for SILVANER, to distinguish it from Petit Rhin, or RIESLING.

🍇 **GROS VERDOT,** undistinguished and unusual Bordeaux variety without the concentration or interest of PETIT VERDOT. May be grown in South America. See VERDOT.

🍇 **GRÜNER VELTLINER,** Austria's most important and very own grape, also grown elsewhere in eastern Europe. In 1992 this well adapted variety was planted on more than a third of Austria's 58,000 ha/143,000 acres of vineyard, particularly in Lower Austria, where it represents more than half of total white grape production, and around Vienna where much of it is sold as very young 'Heurige' wine.

The vine is productive and relatively hardy, but ripens too late for much of northern Europe. Yields of 100 hl/ha (5.7 tons/acre) are often achieved in the least distinguished vineyards of the

Weinviertel in Lower Austria and the resulting wine is inoffensive if unexciting. At its best, arguably in the Wachau and in the hands of some of the most ambitious growers in Vienna, Grüner Veltliner can produce wines which combine both perfume and substance, not unlike some Alsace in style. The wine is typically dry, full, peppery, or spicy. Most examples are best drunk young but the most concentrated examples from the Wachau can take on white burgundy characteristics in bottle.

The variety is also grown just over Lower Austria's northern border in Slovakia, where it is known as Veltlin Zelene or Veltlinske Zelené, and in parts of Hungary as Veltlini.

Roter Veltliner (once planted in California) and, less importantly, **Brauner Veltliner** are grown to a much more limited extent in Lower Austria. The combined plantings of both of these darked-skinned mutations of Grüner Veltliner totalled not much more than 200 ha/500 acres, or 1 per cent of the area planted with Grüner Veltliner, in the early 1990s. See also FRÜHROTER VELTLINER, however.

🍇 **GUARNACCIA**, a strain of GRENACHE local to the island of Ischia off Naples.

🍇 **GUTEDEL**, meaning 'good and noble' in German, is not the most obvious synonym for CHASSELAS today, but Germany still grows more than 1,300 ha/3,200 acres of Weisser Gutedel, almost all in the Markgräflerland region of Baden, where it continues to be popular with growers. A dark-berried form, Roter Gutedel, is also known here.

🍇 **GUTENBORNER**, very minor German crossing bred from Müller-Thurgau × Chasselas Napoleon which has had some fruity success in sheltered sites in England and, unusually for a modern crossing, the Rheingau and Mosel-Saar-Ruwer in Germany, where total plantings had fallen to just 10 ha/25 acres by 1990. Its main attribute is its ability to ripen in cool climates.

H

HANEPOOT, traditional Afrikaans name for South Africa's most planted Muscat, MUSCAT OF ALEXANDRIA.

HARRIAGUE, Uruguayan name for TANNAT inspired by the surname of a Basque pioneer.

🍇 **HÁRSLEVELŰ**, aromatic Hungarian variety producing characteristically spicy wines. This is the variety which brings perfume to the FURMINT grapes which make up the majority of the blend for the famous dessert wine Tokaji. It has larger bunches and smaller grapes than Furmint, but also ripens late and can also be affected by noble rot. It is also widely planted elsewhere in Hungary and produces a range of varietal wines which vary considerably in quality and provenance. Good Hárslevelű is typically a deep green gold, very viscous, full, and powerfully flavoured.

The variety's Hungarian name means 'linden leaf', and translates directly into such synonyms as Lipovina, Lindenblättrige, and Feuille de Tilleul. It makes particularly full bodied wines in Villany in the far south of Hungary and is popularly associated with the village of Debrő in the Mátra Foothills (although much of the wine sold as Debrő i Hárslevelű has been a much less specific and distinguished off dry blend).

The variety is also grown over the border from Hungary's Tokaj region in Slovakia and is even more widely grown in South Africa than Furmint, although the wines it produces here are rarely very distinguished.

HEIDA, HEIDEN, Swiss name for TRAMINER and speciality of Vispertermin.

❦ HELFENSTEINER, famed principally as a parent of the successful German crossing DORNFELDER. It is itself a crossing of FRÜHBURGUNDER × TROLLINGER and is essentially a product of Württemberg, where its ability to ripen earlier than Trollinger is valued. After a small flurry of popularity in the early 1970s, it is declining because of its susceptibility to coulure.

❦❦ HERBEMONT, dark-skinned *aestivalis-cinerea-vinifera* hybrid grape widely planted in Brazil because of its resistance to fungal diseases, and sometimes used to produce white wine.

HERMITAGE, frequent synonym for SYRAH found in Australia but outlawed, for obvious reasons, in Europe. Historic synonym for CINSAUT in South Africa and occasional French Swiss synonym for MARSANNE.

❦ HEROLDREBE, marginal German crossing to which the prolific breeder August Herold of the Weinsberg in Württemberg put his name. This PORTUGIESER × LIMBERGER crossing yields regularly and prolifically, about 140 hl/ha (8 tons/acre), but it ripens so late that it is suitable only for Germany's warmer regions, particularly the Pfalz, where two-thirds of its German total of 220 ha/540 acres were planted in 1991 and are better at providing pink than red wine. Its most useful function was that, like Herold's even less popular HELFENSTEINER, it spawned the promising DORNFELDER.

HONDARRABI, Spanish Basque vine varieties. **Hondarrabi Zuri** is light-berried and more common in Guetaria in Guipúzcoa while the dark-berried **Hondarrabi Beltza** grows in Baquio y Valmaseda in Vizcaya.

🍇 **HUMAGNE BLANC**, rare white grape variety grown in Switzerland's Valais region producing rich, heady wines with a good acid level that are capable of evolution in bottle.

🍇 **HUMAGNE ROUGE**, is even more widely planted than HUMAGNE BLANC. It ripens late and can produce unusual, wild-flavoured, sometimes overtly rustic wines which can respond well to barrel ageing. According to Sloan the vine is the subject of a small revival.

🍇 **HUXELREBE**, early 20th century German crossing that has enjoyed some popularity both in Germany and, on a much smaller scale, England. Although like SCHEUREBE and FABER it was actually bred by Dr Georg Scheu at Alzey, this crossing takes its name from its chief propagator, nurseryman Fritz Huxel. It was bred in 1927 from GUTEDEL (Chasselas) and Courtillier Musqué (which is also an antecedent of the popular hybrid MARÉCHAL FOCH). The crossing is capable of producing enormous quantities of rather ordinary wine—so enormous in fact that the vine's wood can collapse under the strain. If pruned carefully, however, and planted on an average to good site, it can easily reach Auslese must weights even in an ordinary year and produce a fulsome if not exactly subtle wine for reasonably early consumption. Huxelrebe's flavours are more reminiscent of Muscat than Riesling and in England its ripeness is a useful counterbalance to naturally high acidity. In Germany it is grown almost exclusively in the Pfalz and Rheinhessen and, although it is slowly losing ground, there were still more than 1,500 ha/3,700 acres in 1990.

I

INCROCIO, Italian for vine crossing. A wine made from **Incrocio Manzoni** grapes, for example, is made from one of Signor Manzoni's many crossings of one *vinifera* variety with another. His Incrocio Manzoni 6.0.13, RIESLING × PINOT BLANC, is the most widely planted, particularly in north east Italy. Incrocio Manzoni 2.15 is PROSECCO × CABERNET SAUVIGNON.

Incrocio Terzi No 1. is a BARBERA × CABERNET FRANC crossing grown in Lombardy.

🍇 **INZOLIA**, quantitatively important variety grown mainly in Sicily and to a much more limited extent in Tuscany, where it is also known as Ansonica or Anzonica. It was planted on a total of nearly 13,000 ha/32,000 acres of Italian vineyard in the early 1990s, although total plantings were declining. It is grown mainly in western Sicily, where it is valued as a relatively aromatic ingredient, often with the much more common CATARRATTO, in dry white table wines. The best examples show a certain nuttiness, the worst could do with more acid and more flavour.

⚲ **IRSAY OLIVER,** aromatic, relatively recent crossing grown in Slovakia and also Hungary, where it is known as **Irsai Olivér**. This eastern European cross of Pozsony × Pearl of Csaba was originally developed in the 1930s as a table grape. It ripens extremely early and reliably (although it is prone to powdery mildew) and produces relatively heavy, but intensely aromatic wines strongly reminiscent of MUSCAT.

⚲ **ISABELLA, ISABELLE,** widely distributed and widely planted *labrusca × vinifera* American hybrid of unknown origin. It is said to have been named after a southern belle, Mrs Isabella Gibbs, and to have been developed in South Carolina in 1816. It can withstand tropical and semi-tropical conditions and has been planted all over Portugal, Ukraine, Japan, and occasionally crops up in the southern hemisphere, notably in Brazil, where it is by a substantial margin the leading vine. In New York State it was one of the first hybrids to be planted after phylloxera's late 19th century devastation but it has largely been replaced by CONCORD. New plantings were banned in France in 1934. The vine is high yielding but the wines are very obviously foxy.

⚲ **ITALIA,** table grape from which wine is sometimes made in Australia.

ITALIAN RIESLING, or **ITALIAN RIZLING,** sometimes **Italianski Rizling,** all synonyms for WELSCHRIESLING, also known as Riesling Italico.

⚲ **IZSÁKI,** low-quality, late ripening grape, sometimes called White Kadarka, planted in declining quantity on the Hungarian Great Plain.

J

⚲ **JACQUÈRE,** common grape in Savoie in the French alps, where it produces high yields of lightly scented, essentially alpine dry white. Plantings once again increased in the 1980s so that there were 1,000 ha/2,500 acres by 1990.

⚲ **JAÉN,** workhorse variety commonly planted in central and western Spain, and in Portugal's Dão region, where its wines are notable for their lack of acidity. The wine produced, in great quantity, is rustic and undistinguished.

The light-berried **Jaén Blanco** is said by some to be the AVESSO of Portugal.

🍷 **JAMPAL,** southern Portuguese grape with possibly unrealized potential.

🍷 **JOÃO DE SANTARÉM**, name used for the widely planted CASTELÃO FRANCÊS in parts of Portugal's Ribatejo region.

🍷 **JOHANNISBERG RIESLING,** sometimes abbreviated simply to **JR,** common synonym for the great white RIESLING grape of Germany, notably in California. There is no direct connection with the famous Schloss Johannisberg in the Rheingau region except that both the famous castle and the region's reputations are founded on Riesling.

🍷 **JOUBERTIN,** occasionally **JAUBERTIN,** now almost extinct but once widely distributed vine originally from Savoie in south east France. Most notable for its productivity.

🍷 **JUAN GARCIA,** crisp, lively, local speciality of the Fermoselle-Arribes zone west of Toro in north west central Spain, where it is usually mixed in the vineyard with other, lesser vines. A total of about 2,500 ha/6,000 acres are planted and, on rocky hillside sites, the vine can produce highly perfumed if relatively light reds.

🍷 **JUAN IBAÑEZ,** vine grown to a limited extent and mainly in mixed vineyards in Cariñena in north east central Spain. Known as Miguel del Arco in Calatayud.

🍷 **JUHFARK,** distinctive but almost extinct Hungarian vine today found almost exclusively in the Somló region, where it can produce wine usefully high in acidity which ages well. It is usually blended in with the more widely planted FURMINT and RIESLING. The vine, whose name means 'ewe's tail', is inconveniently sensitive to both frost and mildew.

🍷 **JURANÇON,** rare but rather ordinary vine, or rather family of vines, once common in South West France. A light-berried **Jurançon Blanc** also exists.

🍷 **JUWEL,** German crossing of which only about 30 ha were planted in Germany in the early 1990s, mainly in the Rheinhessen.

K

🍷 **KADARKA,** Hungary's most famous red wine grape, largely because of the important role it once played in Bull's Blood. Typically, Kadarka is too often over-produced and picked when still low in colour and flavour and is no longer the backbone of Hungary's red wine production. It has been substantially replaced by the viticulturally sturdier KÉKFRANKOS, and KÉKOPORTO in Villány, and is now Hungary's second most important red grape.

It is still cultivated on the Great Plain and in the Szekszárd wine region just across the Danube to the west but its tendency to grey rot and its habit of ripening riskily late limits it to certain favoured sites. The vine is also naturally highly productive and needs careful control in order to produce truly concentrated wines. Fully ripened Szekszárdi Kadarka can be a fine, tannic, full bodied wine worthy of ageing but is produced in minuscule quantities.

Kadarka's origins are obscure, but some believe it is related to the variety known as Skadarsko, from Lake Scutari, which forms the frontier between Albania and Montenegro.

Today the variety is cultivated on a very limited scale over the eastern border in Neusiedlersee in Austria, over the southern border in Vojvodina in what was Yugoslavia, in Romania, where it is called Cadarca, and, most importantly, in Bulgaria, where it is called Gamza and is widely planted in the north and can produce wines of interest in long growing seasons if yields are restricted.

Because of its, largely historic, fame, this is a variety which is often included in any large nursery collection.

🍇 **KANZLER,** modern German vine crossing already falling from grace. A Müller-Thurgau × Silvaner cross bred at Alzey in 1927 and always essentially a Rheinhessen variety, it reaches high must weights but needs a good site and, most fatally, does not yield well.

KÉK means blue in Hungarian and, as such, can be a direct equivalent of BLAU in German or even Noir in French.

KÉKFRANKOS, Hungarian name for BLAUFRÄNKISCH (of which it is a direct translation). This useful variety, which produces lively, juicy, peppery, well-coloured reds for relatively early consumption is grown widely in Hungary. It is most successful in Sopron near the Austrian border, although it can also produce full bodied wines in Villány. On the Great Plain its wines can be relatively heavy.

🍇 **KÉKNYELŰ,** revered but rare Hungarian vine named after its 'blue' stalk. Once widely planted, it was becoming rare even in its last stronghold Badacsony on the north shore of Lake Balaton in the mid 1990s. The vine itself is so sensitive that yields are extremely low and so it fell from favour in the 1970s and 1980s when wine-making philosophy in Hungary was to produce large quantities of ordinary wine for export to other Comecon countries. True, well-made Kéknyelű can be aromatic and exciting, but some very ordinary blends have been labelled Badacsony Kéknyelű.

🍇 **KÉKOPORTO,** sometimes written **Kékportó,** useful variety grown in Hungary which may be the same as Germany's PORTUGIESER. It produces well-coloured lively red wine not unlike that of KÉKFRANKOS but with a little more body and possibly a better aptitude for cask ageing. It is grown in Eger of Bull's Blood fame, in the red wine region of Villány, where it can yield wines of real concentration, and, with slightly less success, on the Great Plain. Often called simply Oporto, it is also grown in Romania.

🍇 **KERNER,** the great success story of modern German vine-breeding. Bred only in 1969, four or five decades after crossings such as SCHEUREBE, FABER, HUXELREBE, Kerner had almost overtaken the ancient SILVANER to become Germany's third most planted vine by 1990, presumably because it ripens so reliably almost anywhere. As with most 20th century crossings, the bulk of Germany's 8,000 ha / 20,000 acres of Kerner is planted in the Rheinhessen and Pfalz but it is still popular in Württemberg, where it was bred from a red parent TROLLINGER (Schiava Grossa) × RIESLING. The large white berries produce wines commendably close to Riesling in flavour except with their own leafy aroma and very slightly coarser texture. It is a crossing that does not need to be subsumed in the blending vat but can produce fine varietal wines, up to quite high levels of ripeness, on its own account. Of the 20th century *vinifera* crossings, only the more capricious EHRENFELSER is as Riesling-like, both crossings having the ability to age thanks to their high acidity. Kerner is popular with growers as well as wine drinkers because of its late budding and therefore good frost resistance. It is so vigorous, however, that it needs careful summer trimming. It ripens slightly later than Müller-Thurgau, about the same time as Silvaner, but can be planted in almost any vineyard site and regularly achieves must weights and acidity levels 10 to 20 per cent above the dreary Müller-Thurgau.

Kerner, which takes its name not from any vine breeder but from a local 19th century writer of drinking songs, has also been planted in South Africa, but it seems unlikely that there is a long future for it there. A small area of English vineyard is also planted with it.

KEVEDINKA, also known as **Kövidinka,** ordinary white eastern European variety. See DINKA.

KIRÁLYLEÁNYKA, Hungarian name for Romania's grapey FETEASCĂ Regală, a crossing of Fetească Albă and Grasă, or LEÁNYKA. It is sometimes known as Dánosi Leányka after its geographical origins in Transylvania. Balaton Boglar is its chief home in modern Hungary.

KIŞMIŞ, KISMIS, KISHMISH, Middle Eastern synonyms for the common SULTANA.

KLEVNER, like CLEVNER, is, and more particularly was, used fairly indiscriminately in Alsace and other German speaking wine regions for various vine varieties, notably but not exclusively for CHARDONNAY and various members of the PINOT family. References to Klevner or **Klevener** in Alsace in the mid 16th century are common. Today, Klevner and Clevner are most likely to be used for PINOT BLANC.

KNIPPERLÉ

Klevner or **Klevener de Heiligenstein** is an Alsace oddity, wine made in the village of Heiligenstein from the particular strain of the related SAVAGNIN Rosé and GEWÜRZTRAMINER introduced to the village probably from Chiavenna in the Italian alps. It has been a speciality for at least two centuries, although barely 20 ha / 50 acres are planted today.

🍇 **KNIPPERLÉ**, almost a relic, a dark-berried vine once popular as the base of light white wine in Alsace. Occasional bottles can still be found but this subvariety of RÄUSCHLING, relatively popular as a high yielder and early ripener at the end of the 19th century, is too prone to rot to be of more than historical interest.

🍷 **KOTSIFALI**, generous, spicy if soft wines produced from this speciality of the Greek island of Crete, planted on about 550 ha / 1,350 acres. Best blended with something more tannic such as MANDELARI.

KÖVIDINKA, ordinary white eastern European grape variety. See DINKA.

🍷 **KRATOSIJA**, relatively important grape in Macedonia and Montenegro.

KUC. See TRBLJAN.

L

🍷 **LACRIMA (DI MORRO)**, fast maturing, strangely scented speciality of the Marches on Italy's Adriatic coast.
Lacrima Nera is sometimes used as a synonym for GAGLIOPPO.

🍇 **LADO**, high-acid variety grown in Galicia, north west Spain and, particularly, in Ribeiro.

LAFNETSCHA. See COMPLETER.

🍇 **LAGORTHI**, rare Greek variety whose aromatic produce may save it from extinction.

🍷 **LAGREIN**, variety in Trentino-Alto Adige in the Italian Tyrol. Although often over-produced, it can produce Lagrein Scuro or Lagrein Dunkel, velvety reds of real character, as well as fragrant yet sturdy rosés called Lagrein Rosato or Lagrein Kretzer. According to Burton Anderson, this variety, whose name suggests origins in the Lagarina valley of Trentino, was mentioned as early as the 17th century in the records of the Muri Benedictine monastery near Bolzano in Alto Adige.

LAIRÉN, southern Spanish name for AIRÉN.

LAMBRUSCO, central Italian variety grown mainly in the three central provinces of Emilia—Modena, Parma, and Reggio Emilia—although also across the river Po in the province of Mantova, and occasionally as far afield as Piedmont, the Trentino, and even Basilicata. This robust vine, of which there are at least 60 known subvarieties, has been known for its exceptional productivity since classical times.

Modern Lambrusco is a frothing, fruity, typically red (but often decolorized white or pink) wine meant to be drunk young. Wine labelled Lambrusco di Sorbara is made from the subvariety **Lambrusco di Sorbara** and, the most planted, **Lambrusco Salamino**. Wine labelled Lambrusco Grasparossa di Castelvetro must be made mainly from the **Lambrusco Grasparossa di Castelvetro** subvariety. Lambrusco Reggiano is produced principally from the **Lambrusco Marani** and Lambrusco Salamino clones, with **Lambrusco Maestri** and **Lambrusco Montericco** permitted (although they are gradually disappearing). The **Lambrusco Salamino di Santa Croce** clone (whose small bunches are thought to resemble a 'small salami') should form 90 per cent of the synonymous wine. Lambrusco Reggiano on the other hand, the great American success story in the late 1970s and early 1980s, tends to be slightly sweet, the sweetness generally being supplied by the partially fermented must of the ANCELLOTTA grape, which DOC rules permit (up to a maximum of 15 per cent) in the blend.

There are also several hundred ha of a red grape variety known as **Lambrusco Maesini** in Argentina.

LASKI RIZLING, name current in Slovenia, Vojvodina, and some other parts of what was Yugoslavia for WELSCHRIESLING. The vine is cultivated widely in former Yugoslavia, but most successfully in the higher vineyards of Slovenia (just over the border from the spirited Welschrieslings of Styria) and Fruš ka Gora in Vojvodina, where it can produce equally crisp and delicately aromatic wines. Few of these have been exported, however, and most of the large bottling enterprises have been hampered by poor equipment and importers who have been more concerned with quantity than quality. For decades a Slovenian brand, Lutomer Riesling (eventually renamed Lutomer Laski Rizling after German lobbying), was the best selling white wine in the UK, its heavily sweetened style conveying little of the intrinsic character of the variety. See also RIZLING.

LAUZET, almost extinct and not particularly exciting vine theoretically allowed in the wines of Jurançon.

LEÁNYKA, Hungarian name for the white grape variety called FETEASCĂ Albă in Romania. A varietal Leányka has long been produced in Eger, north west Hungary. It can produce good quality wine if yields are restricted. See also KIRÁLYLEÁNYKA.

LEATICO, synonym for ALEATICO.

LEMBERGER

LEMBERGER, Washington State's own name for the lively BLAUFRÄNKISCH vine of central Europe, presumably a corruption of the German synonym Limberger.

🍷 LEN DE L'EL or LEN DE L'ELH, has, like the MANSENGS, been a beneficiary of proud regionalism in South West France. Once a major, it is now a compulsory minor ingredient in the white wines of Gaillac. The wine is powerful and characterful but can be flabby. Its name is local dialect for *loin de l'œil*, or 'far from sight'. This vigorous vine needs a well-ventilated, well-drained site if it is to escape rot in lesser years.

LEXIA, Australian name for MUSCAT OF ALEXANDRIA.

🍷 LIATIKO, ancient Cretan vine producing relatively soft wine, usually blended with the stronger MANDELARIA and KOTSIFALI to make sweet reds, exported in great quantity by Venetian merchants in medieval times. The name suggests the Tuscan vine ALEATICO but there is no ampelographical evidence for a relationship.

🍷 LIMBERGER, also known as **Blauer Limberger** or **Lemberger**, is the German name for the black grape variety much more widely grown in Austria as BLAUFRÄNKISCH. Germany has only about a quarter as much of the variety planted, nearly 700 ha/1,700 acres in the early 1990s, almost exclusively in Württemberg, where both climate and consumers are tolerant of pale reds made from late ripening vines (see TROLLINGER). The wine, often blended with Trollinger to produce a light red suitable for early drinking, has a better colour than that of most Germanic red wine varieties and has a good bite, notably of acidity. Washington State, curiously, has sizeable plantings of this variety, known there as Lemberger.

🍷 LIMNIO, ancient vine native to the island of Lemnos in Greece, where it can still be found. It has also transferred successfully to Halkidiki in north east Greece, however, where it produces a full bodied wine with a good level of acidity. It is thought that this is the variety which Aristotle called Lemnia.

LINDENBLÄTTRIGE, German synonym for Hungary's light-berried HÁRSLEVELŰ.

LIPOVINA, Czech synonym for Hungary's light-berried HÁRSLEVELŰ.

🍷 LISTAN, synonym for PALOMINO, the grape that can produce superb sherry around Jerez in southern Spain, but results in dull, flabby white table wines almost everywhere else. Listan is the name by which the variety is known in much of Spain and in France. There are still several hundred ha of it in the western Languedoc and in armagnac country but it is being systematically grubbed up. Crossed with Chardonnay, it spawned CHASAN.

🍇 **LISTAN NEGRO**, recently appreciated grape which dominates wine production on the Spanish island of Tenerife, planted on more than 5,000 ha / 12,400 acres. Carbonic maceration has managed to coax exceptional aromas out of this medium bodied wine. The grape may also be called Almuñeco.

🍇 **LLADONER PELUT, LLEDONER PELUT,** Catalonian name for the downy-leaved form of GRENACHE, Grenache Poilu or Velu. Both vine and wine closely resemble Grenache Noir except that the underside of the leaves of the Lladoner Pelut are downier. It is officially and widely sanctioned in Languedoc-Roussillon appellation wines, often being specified alongside Grenache, and has the advantage of being less susceptible to rot.

In practice it is declining in popularity but in the early 1990s there were more than 150 ha planted with the variety in the Aude and Roussillon as well as some plantings in Spain around Tarragona.

🍇 **LOUREIRO**, fine, 'laurel-scented' variety grown in Vinho Verde country in northern Portugal and also, increasingly, as **Loureira** in Rias Baixas, the Galician region in north west Spain. Plantings probably total nearly 3,000 ha / 7,400 acres. It has often been blended with TRAJADURA but can also be found in aromatic varietal form. It can yield quite productively in the north of the Vinho Verde region.

LUNEL, occasional Hungarian name for a yellow-berried form of MUSCAT BLANC À PETITS GRAINS grown in the Tokaj region.

M

🍇 **MACABEO, MACCABÉO, MACCABEU,** northern Spain's most popular light-skinned grape and also extremely widely planted over the Pyrenees in Roussillon and the Languedoc, where it plays a part, often blended with BOURBOULENC and/or GRENACHE BLANC, in most appellation white wines and many of the strong sweet specialities of Roussillon. Macabeo spread to southern France from Spain but Odart claims that its origins are Middle Eastern. It is a vigorous vine that buds conveniently late for regions prone to spring frosts and can be quite productive so long as autumns are dry and the possibility of rot is minimized. Well established at one time in North Africa, the vine can tolerate hot, dry conditions.

The wine produced tends to have a vaguely floral character and relatively low acidity unless the grapes are picked so early that the floral character is even more difficult to discern, but it has the advantage of withstanding oxidation well, unlike the Grenache Blanc with which it is often grown.

Perhaps this is one of the reasons why it has been so enthusiastically embraced by the growers of Rioja where, as Viura, it has all but displaced Malvasia and Garnacha Blanca (Grenache Blanc) to represent more than 90 per cent of all white varieties planted. The fact that it is so much better suited to making light whites for early consumption than heavy, oak-aged wines for the long term may help explain the stylistic evolution of white Rioja.

It is also grown widely in Penedès and, especially, Conca de Barberá where, with Parellada and Xarel-lo, it makes up the triumvirate of varieties for sparkling Cava, as well as being found throughout north-eastern Spain as far south as Tarragona.

🍇 **MACERATINO**, increasingly rare grape on the Adriatic coast of Italy, possibly related either to GRECO or the local VERDICCHIO, according to Anderson.

MÄDCHENTRAUBE, German synonym for FETEASCA.

🍇 **MADELEINE ANGEVINE**, table grape popular with English vine-growers for its ability to ripen early, although several different, if related, vines may have been imported under this name. Can make some attractive, curranty whites.

🍷 **MAGARATCH BASTARDO**, a fortified wine variety created at the Magaratch wine research institute in the Crimea by crossing the Portuguese BASTARDO with SAPERAVI.

🍷 **MAGARATCH RUBY,** Crimean crossing of CABERNET SAUVIGNON and SAPERAVI achieved at the Magaratch institute. A newer generation of varieties has been developed with specific resistances to various pests and diseases.

🍷 **MAGLIOCCO CANINO**, speciality of Calabria in southern Italy where it is commonly blended with GAGLIOPPO, with which Anderson posits a relationship. There were just over 1,500ha/3,700 acres in 1990.

🍇 **MAJARCĂ ALBĂ**, Romanian speciality.

MALA DINKA, occasional Bulgarian name for GEWÜRZTRAMINER.

🍇 **MALAGOUSIA, MALAGOUSSIA**, elegant grape variety rediscovered and identified only recently in modern Greece. It may be related to MALVASIA and yields similarly full bodied, perfumed wines.

🍷 **MALBEC**, variety once popular in Bordeaux but now more readily associated with Argentina and Cahors in South West France, in both of which it is the most planted vine. It has many synonyms (Galet cites Côt as its true name) and is known in much of western France, including the Loire, where it is quite widely grown. In Argentina it is often called **Malbeck.** On the north bank of Bordeaux's Gironde it is called Pressac; in Cahors, suggesting origins in northern Burgundy, Auxerrois. Galet's complete list of synonyms runs to nearly 400 words, however, for the variety has at one time been grown in 30 different *départements* of France.

Malbec has been declining in popularity in France for it has many of the disadvantages of Merlot (sensitivity to coulure, frost, downy mildew, and rot) without as much obvious fruit quality. Indeed it can taste like a rather rustic, even shorter-lived version of Merlot, although when grown on the least fertile, high, rugged limestone vineyards of Cahors it can occasionally remind us why the English used to refer to Cahors as 'the black wine'. Cahors appellation contrôlée regulations stipulate that 'Cot' must constitute at least 70 per cent of the wine. Other south western appellations in which Malbec may play a (smaller) part are Bergerac, Buzet, Côtes de Duras, Côtes du Frontonnais, Côtes du Marmandais, Pécharmant, and Côtes du Brulhois. It is also theoretically allowed into the Midi threshold appellations of Cabardès and Côtes de la Malepère, but is rarely found this far from Atlantic influence.

At one time, especially before the predations of the 1956 frosts, Malbec was quite popular in Bordeaux and is still permitted by all major red bordeaux appellations, but total plantings fell from 4,900 ha/12,000 acres in 1968 to 1,500 ha in 1988. It persists most obviously in Bourg, Blaye, and the Entre-Deux-Mers region but is being replaced with varieties whose wines are more durable.

Blended with Cabernet and Gamay, it is also theoretically allowed in a wide range of mid-Loire appellations—Anjou, Coteaux du Loir, Touraines of various sorts, and even sparkling Saumur—but has largely been replaced by Cabernets Franc and Sauvignon.

There is a tiny planting of it in Spain's Ribera del Duero region, but it is in Argentina where the variety really holds sway, and has been the most planted dark-skinned variety for some time. Varietal Argentinian Malbec can have some perceptibly Bordelais characteristics, of flavour rather than structure, and the wines have a luscious, gamey concentration and ageing potential, especially in the Lujan de Cuyo region of Mendoza, quite unknown elsewhere in the world. Malbec, sometimes called Cot, is also important in Chile (though not as important as Pais and Cabernet Sauvignon) and there were about 4,000 ha/10,000 acres of it planted in the early 1990s. Chile's version tends to be more tannic than those raised across the Andes.

Australians have no great respect for their Malbec and have been uprooting it systematically, but still had 250 ha in 1990. Californians had got their total, significant before Prohibition, down to about the same level, but who knows when this may be driven up by a desire to replicate, for example, the country wines of France. Much of today's California Malbec is dutifully added to Meritage wines blended in the image of Bordeaux.

A small amount of Malbec, Malbech, or Malbeck is also planted in north east Italy, as it is in South Africa.

MALI PLAVAC

MALI PLAVAC. See PLAVAC MALI.

🍷 MALVAR, grape commonly grown around Madrid producing slightly rustic wines but with more body and personality than its ubiquitous neighbour AIRÉN. Plantings totalled about 2,500 ha/6,200 acres in the mid 1990s.

🍷🍷 MALVASIA, name used widely, especially in Italy and Iberia, for a complex web of grape varieties, typically ancient and of Greek origin and producing characterful wines high in alcohol and, often, sweetness. Most are deeply coloured whites but some are, usually light, reds. Malvasia is widely disseminated, and its total plantings make it one of the world's dozen or so most planted light-skinned grapes.

Malvasia is the Italian corruption of Monemvasia, the southern Greek port which, in the Middle Ages, was a busy and natural entrepôt for the rich and highly prized dessert wines of the eastern Mediterranean, notably those of Crete, then called Candia. **Malvasia di Candia** is today one distinctive subvariety of Malvasia which, like the somewhat similar Muscat, exists in many guises and hues. So important was Malvasia during the time of the Venetian republic that Venetian wine shops were called *malvasie*.

The French corruption of Malvasia has been used particularly loosely; for more details see MALVOISIE. The word was also corrupted into Malmsey in English, which continues to be an important style of madeira, traditionally based on the Malvasia grape. The Germans call their various though rare forms of Malvasia **Malvasier** and occasionally early, or *früh*, VELTLINER in various colours of berry.

Malvasia, in at least 10 distinctive and various forms—white and red, dry and sweet—is one of Italy's most widely planted grapes, with close to 50,000 ha/123,000 acres being cultivated in the early 1990s in regions as distant and disparate as the Basilicata and Piedmont. One of the most quantitatively important was the **Malvasia Bianca di Chianti** or **Malvasia Toscana** of Tuscany, grown throughout Latium and Umbria too, although its common blending partner TREBBIANO is very much more popular with growers for its productivity. Malvasias in general are slightly prone to oxidation, which is not a major disadvantage in the production of vin santo, to which Malvasia is particularly well suited. Malvasia, often Malvasia di Candia, and Trebbiano are the classic ingredients in all sorts of central Italian whites such as Frascati, Marino, and Est!Est!!Est!!! This type of wine has lost considerable ground in Tuscany, and then Umbria, since the 1970s, however, as producers have replaced both Malvasia and Trebbiano with more strongly characterized international vine varieties. Monovarietal Malvasia wines are rare in central Italy, but pioneering efforts began to appear in the Castelli Romani zone in the early 1990s. In Piedmont in the far north west, a dark, sweet red Malvasia is produced near Asti in the Casorzo d'Asti and Castelnuovo Don Bosco zones usually from **Malvasia di Casorzo**, although a **Malvasia di Schierano** is also known.

The finest Italian dry white varietal Malvasia is made in Friuli, where two DOCs—Collio and Isonzo—cultivate what is called locally **Malvasia Istriana** and which, according to local tradition, was carried to these vineyards from Greece by medieval Venetian seafarers. A lightly sparkling Malvasia is also produced in the Colli Piacenti and parts of Emilia.

Sweet white Malvasia, made from dried grapes and once considered one of Italy's finest dessert wines, has fallen from favour so that many traditional Malvasia wines are now practically extinct. **Malvasia delle Lipari** seemed destined to share their fate but was revived in the 1980s and the survival of this distinctive sweet orange relic from the volcanic island of Lipari off Sicily seems assured. The Basilicata has its own version of sweet Malvasia, produced in the same zone as Aglianico del Vulture, which also exists in dry and spumante versions. Sardinia has its own **Malvasia Sarda**.

Dark-skinned **Malvasia Nera**, particularly common in Alto Adige, is most commonly used in conjunction with other grapes: as the minority partner of NEGROAMARO in the standard red blend of the provinces of Lecce and Brindisi in Puglia, and as a useful supplement to Sangiovese in Tuscany, where it adds both colour and perfume. The introduction of the even more aromatic and deeply coloured Cabernet Sauvignon to Sangiovese-based Tuscan blends in the 1970s and 1980s has led to a distinct loss of favour for Malvasia Nera and uncertain long-term prospects.

Piedmont is the only significant producer of varietal Malvasia Nera wines, with two DOC zones: Malvasia di Casorzo, in both a dry and sweet version, and **Malvasia di Castelnuovo Don Bosco**. Total area planted is less than 100 ha / 250 acres, however, and total annual production less than 4,000 hl / 106,000 gal.

On the French island of Corsica, most growers believe that their Malvoisie is identical to VERMENTINO, which may be related to the greater Malvasia family.

Malvasia is planted to a declining extent in northern Spain, notably in Rioja and Navarra, although the less interesting Viura, or MACABEO, has been gaining ground. Malvasia is also planted in Valencia, Zamora, and on the Canary Islands.

Also well into the Atlantic, **Malvasia di Candida** in particular is beginning to gain ground once more in the vineyards of Madeira, where it was the variety most commonly used to produce Malmsey, before the advent of phylloxera.

Myriad Malvasias are also grown on mainland Portugal, making a contribution to such varied wines as Buçaco, Colares, and, as the distinctly ordinary **Malvasia Rei** (also known as Seminario), it is an ingredient in white port.

In California more than 2,000 acres / 800 ha of **Malvasia Bianca,** the most substantial plantings being in Tulare county at the very southern end of the Central Valley, are enjoying new-found fame, benefiting from the current fashion for all things Italian and producing tangy off-dry whites with real substance and character.

The variety is grown to a very limited extent in modern Greece, mainly on the Aegean islands of Paros and Syros but **Malvazia** is the third most planted white wine grape in the former Yugoslavia.

MALVOISIE, is one of France's most confusing vine names, perhaps because, like PINEAU, the term was once used widely as a general term for superior wines, notably those whose origins were supposed to be Greek. There is no single variety whose principal name is Malvoisie, but it has been used as a synonym for a wide range of usually light-berried grape varieties producing full bodied, aromatic whites. Despite the etymological similarity, Malvoisie has rarely been a synonym for the famous MALVASIA of Greece and Madeira. Malvoisie is today found on the labels of some Loire and Savoie wines made from such plantings of PINOT GRIS as remain, just as **Malvoisie du Valais** is a common synonym for Pinot Gris in Switzerland. It is also sometimes used for BOURBOULENC in the Languedoc, and occasionally for MACCABEU in the Aude, for CLAIRETTE in Bordeaux, and for TORBATO in Roussillon. VERMENTINO, which may in fact belong to the Malvasia family, is sometimes called Malvoisie in Iberia and is known as **Malvoisie de Corse** in Corsica.

Malvoisie Rose and **Malvoisie Rouge** are occasionally used as synonyms for FRÜHROTER VELTLINER in Savoie and northern Italy, while the **Malvoisie Noire** of the Lot in South West France may be TROUSSEAU.

🍇 **MAMMOLO**, perfumed red grape variety producing light wines which supposedly smell of violets, or *mammole*, in central Italy. The variety, a permitted ingredient in Chianti, is relatively rare today, although a small amount is also grown in the Vino Nobile di Montepulciano zone. The 1990 Italian vineyard census found 61 ha/152 acres in total. Its significance, as an often theoretical seasoning to the blend, may be compared to that of PETIT VERDOT in the classic Médoc of Bordeaux.

🍇 **MANDELARI, MANDELARIA**, powerful speciality of various Greek islands, including Crete, where it is often blended with the much softer KOTSIFALI. Probably Greece's third most planted red wine variety. Grapes have thick skins and therefore wine produced is deep coloured and notably high in tannins. It can produce harmonious dry reds such as Peza, or even sweet reds. The AMORGHIANO of Rhodes is probably related.

🍇 **MANSENG**, in both GROS MANSENG and, finer, PETIT MANSENG forms is the classic Basque variety responsible for the exceptional tangy-rich white wines of South West France. Such has been Jurançon's return to public favour, and corresponding Gascon enthusiasm for Pacherenc, that Manseng was one of the few grape arieties to increase its hold on French viticulture in recent years, rom just 90 ha in 1968 to 1,152 ha/2,845 acres two decades later although this apparent increase may be due more to Gascony's ecent viticultural archive work than to new plantings).

Manseng, usually Petit Manseng, can also be found in Uruguay vhere it, like TANNAT, was taken by Basque settlers in the 19th entury. Petit Manseng is also increasingly being planted by such ine variety enthusiasts as Aimé Guibert of Mas de Daumas Gassac in the Languedoc and Randall Grahm of Bonny Doon in California.

MANTONEGRO, principal Majorcan grape, producing scented out light wines which tend to age and oxidize fast. Best blended vith a more structured grape.

MANTONICO BIANCO, ancient vine, probably of Greek origin, grown on just over 1,100 ha/2,700 acres in Calabria. A **Mantonico Nero** also exists.

MARÉCHAL FOCH, French hybrid Kuhlmann 188.2 named after a famous French First World War general, bred by Kuhlmann of Alsace from a *riparia-rupestris* hybrid and the *vinifera* variety Goldriesling (RIESLING × Courtiller Musqué, which Galet claims s a particularly early ripening MUSCAT). It has good winter hardiness and ripens very early. It was once widely cultivated in the Loire and is still popular in Canada and New York State where t is spelt **Marechal Foch** and may be made by carbonic maceration. It produces fruity, non-foxy wines with a very loose and much-promoted similarity to PINOT NOIR. Although it is sometimes given oak ageing, the wine is not particularly stable.

MARIA GOMES, Bairrada synonym for the Portuguese white grape variety FERNÃO PIRES.

MARIA ORDOÑA, See MERENZAO.

MARQUÉS, occasional name for LOUREIRO.

MARSANNE, is an increasingly popular white grape variety probably originating in the northern Rhône, where it has all but taken over from its traditional blending partner ROUSSANNE in such appellations as St-Joseph, St-Péray, Crozes-Hermitage, and, to a slightly lesser extent, Hermitage itself. The vine's relative productivity has doubtless been a factor in its popularity, and modern wine-making techniques have helped mitigate Marsanne's tendency to flab. It is increasingly planted in the Midi where, as well as being embraced as an ingredient in most appellations, it is earning itself a reputation as a full bodied, characterful varietal, or a blending partner for more aromatic, acid varieties such as Roussanne, VIOGNIER, and Rolle (VERMENTINO). The wine is particularly deep coloured, full bodied with a heady, if often heavy, aroma of glue verging occasionally on almonds. It is not one of the chosen varieties for Châteauneuf-du-Pape, in which CLAIRETTE shares many of Marsanne's characteristics.

Australia has some of the world's oldest Marsanne vineyards, notably in the state of Victoria, and a fine tradition of valuing this Rhône import and hefty wines it produces, which can brown relatively fast in bottle. In Switzerland, as Ermitage Blanc, it produces a lighter wine that is nevertheless one of the Valais's heaviest.

MARZEMINA BIANCA, occasional southern Italian synonym for CHASSELAS.

♀ **MARZEMINO,** interesting, late ripening variety grown to a strictly limited extent in Trentino and Lombardy in northern Italy. Once much more famous than now, especially in Chianti, it has poor resistance to fungal diseases, and is often allowed to over-produce, but it can yield lively wines, some of them lightly sparkling, which appeal particularly to opera enthusiasts.

MATARO, common synonym of MOURVÈDRE used primarily in Australia, sometimes in Roussillon and, by those who do not realize how fashionable Mourvèdre has become, in California. It is almost invariably the case that someone who refers to Mourvèdre as Mataro does not have a particularly high opinion of it.

♀ **MATRASSA,** dominant grape of Azerbaijan between Armenia and the Caspian Sea. It is also found further east in the Central Asian republics of the CIS. Also called Kara Shirei and Kara Shirai.

♀ **MAUZAC,** or more properly **Mauzac Blanc,** a declining but still surprisingly important white grape in South West France, especially in Gaillac and Limoux where it is the traditional and still principal vine variety. It produces relatively aromatic wines, often tasting of lightly dried apple peel, which are usually blended, with LEN DE L'EL around Gaillac and with CHENIN BLANC and CHARDONNAY in Limoux. In the 1980s plantings declined in Gaillac and rosé in Limoux to total nearly 6,000 ha / 15,000 acres in France by the end of the decade.

The vine, whose yields can vary enormously according to site, buds and ripens late and grapes were traditionally picked well into autumn so that musts fermented slowly and gently in the cool Limoux winters, ready to referment in bottle in the spring. Today Mauzac tends to be picked much earlier, preserving its naturally high acidity but sacrificing much of its shrivelled apple flavour, before being subjected to the usual champagne-making technique. In Gaillac some Mauzac is still made in various degrees of sweetness and fizziness.

♀ **MAVRO,** means 'black' in Greek and is the common name of the dominant but undistinguished dark-skinned grape variety on the Mediterranean island of Cyprus.

🍇 **MAVRODAPHNE**, Greek vine whose name means 'black laurel' and which is grown mainly around Patra. This aromatic, powerful variety, also grown to a limited extent on the island of Cephalonia, is occasionally vinified dry but only for use as a blending component. Most of it goes into strong, sweet wines made in the port mould, usually the result of extended cask maturation.

MAVRO NEMEAS, alternative name for the dominant Nemean dark-skinned variety AGHIORGHITIKO.

🍇 **MAVROUDI**, almost extinct Greek vine. Same as MAVRUD?

🍇 **MAVRUD**, indigenous Balkan variety most closely associated with Bulgaria. It is capable of producing intense, tannic wine if allowed to ripen fully and is grown exclusively in central southern Bulgaria. A speciality of Assenovgrad near Plovdiv, it is small-berried and low yielding and is usually grown in untidily straggling bushes. The robust wine produced responds well to oak ageing, although it tends to age rather faster than Bulgaria's other noble indigenous vine MELNIK. Mavrud is also grown in Albania.

MAZUELO, MAZUELA, Riojan name for CARIGNAN.

MÉDOC NOIR, Hungarian name for MERLOT.

🍇 **MELNIK**, abbreviated name for the powerful indigenous Bulgarian variety that is grown exclusively around the ancient town of Melnik close to the Greek border in what was Thrace. It may therefore have been cultivated here for many centuries and its wines certainly taste more Greek in their extract, tannin, and alcohol than typical of modern Bulgaria. Its full name is Shiroka Melnishka Losa, or 'broad leaved vine of Melnik', and its berries are notably small with thick, blue skins. Some wines have the aroma of tobacco leaves, another local crop. Oak ageing and several years in bottle bring out a warmth, and powerful subtlety not unlike a Châteauneuf-du-Pape. This is probably the Bulgarian wine with the greatest longevity, but see also MAVRUD.

🍇 **MELON**, or **MELON DE BOURGOGNE**, French variety famous in only one respect and one region, Muscadet. As its full name suggests, its origins are Burgundian, Melon having been outlawed just like GAMAY at various times during the 16th and 17th centuries. Unlike its fellow white burgundian CHARDONNAY, several of whose synonyms include the word Melon, it is not a noble grape variety but it does resist cold well and produces quite regularly and generously. It had spread as far as Anjou in the Middle Ages according to Bouchard and so it was natural that the vine-growers of the Muscadet region to the west might try it. It became the dominant vine variety of the Loire-Atlantique (the mouth of the Loire) in the 17th century when Dutch traders encouraged production of high volumes of relatively neutral white wine, in place of the thin reds for which the region had previously been known, as base wines for Holland's enthusiastic distillers.

Melon's increasing importance today rests solely on Muscadet, whose main attribute could be said to be that it has so few attributes. Indeed, the most successful Muscadet makers do their best to imbue the variety with character by leaving the young wine on its lees (*sur lie*) or gently oaking it. Among France's six most planted white grape varieties, only Chardonnay, Sauvignon Blanc, and Melon increased their total area in the surplus-conscious 1980s, to more than 11,000 ha/27,000 acres in the case of Melon, exclusively around the mouth of the Loire.

Many of the older cuttings of the variety called PINOT BLANC in California are in fact Melon.

♥ MENCÍA, name given to two distinct varieties grown so widely in north west Spain that their total area is about 9,500 ha/23,500 acres in such zones as Ribeira Sacra, El Bierzo, Rias Baixas, Valdeorras, and León. True, indigenous Mencía produces light, pale, relatively fragrant red wines for early consumption while a local strain of CABERNET FRANC introduced in Galicia in the 19th century also, confusingly, goes by this name.

♥ MENU PINEAU, synonym for ARBOIS.

♥ MERENZAO, lesser speciality of the Valdeorras region in north west Spain, sometimes known as Maria Ordoña. Some recent experiments suggest it has real potential. It is sometimes known as Bastardo, but there are no indications it is the same as the Portuguese variety of the same name.

♥ MERILLE, once widely planted south east of Bordeaux. Now a few hundred ha produce undistinguished red wine not permitted in local appellations such as Buzet or Côtes du Marmandais.

♥ MERLOT or **MERLOT NOIR,** variety popularly associated with the great wines of St-Emilion and Pomerol, and Bordeaux's most planted vine by far, which has been enjoying unaccustomed popularity elsewhere. Merlot was already documented as a good quality vine variety in the St-Émilion and Pomerol region in 1784 according to the historian Enjalbert. Merlot's flavour can vary from opulently plummy and fruitcake-like in St-Émilion to a gentler variation on the Cabernet theme, but its texture is almost invariably less tannic and fuller bodied.

Throughout South West France and, increasingly, much of the rest of the world, Merlot plays the role of constant companion to the more austere, aristocratic, long-living CABERNET SAUVIGNON. It early maturing, plump, beguiling fruitiness provides a more obvious complement to Cabernet Sauvignon's attributes than the CABERNET FRANC that often makes up the third ingredient in the common 'Bordelais' (actually Médoc) blend. It also provides good viticultural insurance in more marginal climates as it buds, flowers, and ripens at least a week before Cabernet Sauvignon (although this makes Merlot more sensitive to frost, as was shown dramatically in Bordeaux in both 1956 and 1991). Its early flowering makes it particularly sensitive to coulure, which weaker rootstocks can help to prevent. Merlot is not quite so vigorous as Cabernet Sauvignon but its looser bunches of larger, notably thinner-skinned grapes are much more prone to rot. It is also more sensitive to downy mildew. (Spraying can be a particularly frequent phenomenon in the vineyards of Bordeaux.) Merlot responds much better than Cabernet Sauvignon to damp, cool soils, such as those of St-Émilion and Pomerol, that retain their moisture well and allow the grapes to reach full size. In very well-drained soils, dry summers can leave the grapes undeveloped.

For the vine-grower in anything cooler than a warm or hot climate, Merlot is much easier to ripen than Cabernet Sauvignon, and has the further advantage of yielding a little higher to boot. It is not surprising therefore that in France and northern Italy, total Merlot plantings have for long been greatly superior to those of Cabernet Sauvignon. This is particularly marked in Bordeaux, where Cabernet Sauvignon dominates Merlot only in the famously well-drained soils of the Médoc. Elsewhere, not just in St-Émilion and Pomerol but also in Graves, Bourg, Blaye, Fronsac, and, importantly, those areas qualifying for basic Bordeaux or the rest of the so-called Bordeaux Côtes appellation, Merlot predominates. In 1990 Merlot plantings in the Bordeaux region totalled 44,000 ha / 109,000 acres while those of Cabernet Sauvignon and Cabernet Franc were 25,000 ha and 13,000 ha.

Such was the increase in popularity of Merlot with the vine-growers of Bordeaux, Bergerac, and Languedoc-Roussillon in the 1980s (often switching from less glamorous white wine grapes) that Merlot was France's third most planted black grape variety, after CARIGNAN and GRENACHE, by 1988. Merlot is more widely planted than either sort of Cabernet, not just in Bordeaux but also in the rest of South West France. Wherever in this quarter of France the appellation contrôlée regulations sanction CABERNET SAUVIGNON they also sanction Merlot, although the latter is favoured in the Dordogne while the Cabernets are preferred in Gascony.

With SYRAH, Merlot has been a major beneficiary of the Midi's turn towards 'improving' grape varieties. There were 15,600 ha / 38,600 acres of Merlot in the Languedoc-Roussillon by 1993, much of it sold as a varietal or blended Vin de Pays whose quality is usually in inverse proportion to yield. The only appellations of the Midi to sanction Merlot within their regulations are Cabardès and Côtes de la Malepère.

Merlot is also extremely important in Italy, where there were more than 30,000 ha in 1990. The variety is grown particularly in the north east, often alongside Cabernet Franc, where output of the wine called there 'Merlott' can be easily 100,000 hl (2.6 million gal) a year from the plains of both Grave del Friuli and Piave, even if better, more concentrated wines come in smaller quantities from higher vineyards. In Friuli indeed, where Merlot performs perceptibly better than most Cabernet, there is even a Strada del Merlot, a tourist route along the Isonzo river. Individual denominations for Merlot abound in Friuli, the Veneto, and Trentino-Alto Adige. Merlot is also planted on the Colli Bolognesi in Emilia-Romagna. The variety is planted in 14 of Italy's 20 regions. In general, little is expected from or delivered by the sea of light, vaguely fruity Merlot from northern Italy, which makes it all the more remarkable that the variety is being taken seriously by a handful of producers in Tuscany and Umbria. Ornellaia and the Fattoria de Ama were the first to show that Italy could provide something more in the mould of serious Pomerol.

In Spain total plantings of Merlot for quality wines were only just over 400 ha/1,000 acres in the early 1990s, most of them in Penedès.

The variety is vital to the wine industry of Italian Switzerland, however, and is made at a wide range of quality levels. Merlot has also been popular over Italy's north eastern border in Slovenia and all down the Dalmatian coast, where it can be attractively plummy when yields are restricted. As 'Médoc Noir' it is also known in Hungary, notably around Eger in the north east and Villány in the south. It is also the most widely planted red wine variety in Romania, where there were 11,400 ha/28,200 acres in 1993 and is the second most planted variety in Bulgaria after Cabernet Sauvignon, with which it is often blended. It is also planted in Russia and, particularly, Moldova.

Outside these traditional strongholds, Merlot was until recently taken up much more slowly than the world-famous Cabernet Sauvignon. The fact that it is slightly lower in acidity as well as international acclaim may have hindered its progress in some warmer climates such as Iberia and most of the eastern Mediterranean, where it was still relatively rare in the mid 1990s (although the mid Portuguese Ma Partilha already showed promise).

A lift in Merlot's reputation was already apparent, however, by 1990, most obviously in North America. Merlot was suddenly regarded as the hot varietal in the Cabernet-soaked, fashion-conscious state of California. Although in 1985 California had a total of hardly 2,000 acres/800 ha of Merlot, this had already risen to nearly 18,000 acres/7,300 ha by 1994 and the variety is still in great demand both for blending with other Bordeaux varieties and for varietal reds that were softer, milder, and easier to drink young than the state's Cabernet Sauvignons. Some would describe Merlot as the new Chardonnay: easy to drink but with the reputed health benefits of red wine.

Merlot has had little success in Oregon's vineyards, where the much cooler climate makes coulure too grave a problem, but in Washington's sunny inland Columbia Basin it has produced consistently fine, fruity, well-structured reds. Merlot is the state's most popular black grape variety with over 1,500 acres by 1991, an increase of more than 100 per cent in just three years. Merlot is also grown increasingly in other North American states and has demonstrated a particular affinity with the conditions of Long Island in the state of New York.

In South America Merlot is increasingly important to the wine industry of Chile, which has shown a firm hand in turning out low-cost answers to the fashionable Merlots of California, and in Argentina, where its 2,500 ha/6,000 acres of Merlot in 1989, mostly in Mendoza, was only slightly less than the country's total area of Cabernet Sauvignon—and very much less than that of Malbec.

California's relatively late enthusiasm for Merlot was mirrored in Australia where, of the 500 ha planted in 1990, only 300 were old enough to bear fruit. The great potential for Merlot in selected spots there is slowly being realized, but the prevailing tradition has been to soften Cabernet Sauvignon (inasmuch as ultraripe Australian Cabernet needs softening) with Shiraz.

Merlot clearly has potential in New Zealand too and there were 150 ha planted, mainly for filling in the flavour holes of the more angular Cabernet Sauvignon, in 1992. South Africa has produced some interesting varietal Merlots as well as using it to good effect in Bordeaux blends, but the variety has yet to establish a distinct identity for itself on the Cape.

As the world's wine consumers search for yet more fine red wine that offers an alternative to the unfashionably rigorous charms of Cabernet Sauvignon, it seems likely that Merlot will become increasingly important in countries other than France.

Merlot Noir has no direct relationship to the much less distinguished white-berried **Merlot Blanc** which has also been cultivated, on a much smaller and decreasing scale, in Bordeaux.

🍇 **MERSEGUERA,** lacklustre Spanish variety (Exquitxagos in Pendès) widely grown in Alicante, Jumilla, and Valencia where there are about 19,500 ha/48,000 acres. The vine ripens relatively early and has compact bunches of large grapes.

🍇 **MESLIER ST-FRANÇOIS,** like ARBOIS, a speciality of the Loir-et-Cher *département* in the westward bend of the Loire. It has been disappearing at an even faster rate than Arbois, however. At one time it was used for producing cognac and, especially, armagnac.

🍇 **MEUNIER,** one of France's dozen most planted black grape varieties (even though it and its common synonym Pinot Meunier are hardly ever seen on a wine label). Meunier was probably an early, particularly downy, mutation of the famously mutable PINOT NOIR. It earns its name (*meunier* is French for miller) because the underside of its downy leaves can look as though they have been dusted with flour. In Germany it is known as Müllerrebe (miller's grape) as well as Schwarzriesling.

Meunier is treasured in Champagne, as it was in the once-extensive vineyards of northern France, because it buds later and ripens earlier than the inconveniently early budding Pinot Noir and is therefore much less prone to coulure and more dependably productive. Acid levels are slightly higher although alcohol levels are by no means necessarily lower than those of Pinot Noir. Meunier is therefore the most popular choice for Champagne's growers, especially those in cooler north-facing vineyards, in the damp, frost-prone Vallée de la Marne, and in the cold valleys of the Aisne *département*. Overall, it is planted on more than 40 per cent of the region's vineyards. Plantings of Pinot Noir and Chardonnay increased at a greater rate than those of Meunier in the 1980s, however, partly because of premiums paid for these 'nobler' varieties.

Common wisdom has it that as an ingredient in the traditional three-variety champagne blend, Meunier contributes youthful fruitiness to complement Pinot Noir's weight and Chardonnay's finesse. Although the top drawer house of Krug is publicly enthusiastic about Meunier, few other producers boast of it, and few preponderantly Meunier growers' champagnes have any weight or staying power. Meunier is generally lower in pigments than Pinot Noir, and one of its common French synonyms is Gris Meunier.

It has largely disappeared elsewhere in northern France, although it is still technically allowed into the rosés and light reds of Côtes de Toul, Moselle, Touraine, and Orléanais in the Loire valley.

As Müllerrebe or Schwarzriesling, a selection of Meunier is relatively, and increasingly, popular in Germany, where the majority of its nearly 2,000 ha / 5,000 acres are grown in the Württemberg region. It is also grown in German-speaking Switzerland, and to a much lesser extent in Austria and former Yugoslavia.

Curiously, in Australia Meunier has a longer documented history as a still red varietal wine (at one time called Miller's Burgundy) than Pinot Noir. New-found enthusiasm for authentic replicas of champagne saved the variety from extinction in Australia and the odd, juicy varietal still red can also be found.

It was also with an eye to producing 'genuine' replicas of champagne that growers in California sought Meunier cuttings in the 1980s so that the state's total acreage of the variety was 275 (110 ha) by 1992, almost exclusively in Carneros.

⚘ **MEZESFEHÉR**, speciality of Hungary where its soft, usually sweet, white wines are much admired. Such wines are rarely exported to western markets, however. Its name means 'white honey' and it is perhaps most successful around Eger and Gyöngyös.

🍇 **MILGRANET**, rare speciality of French vineyards north and west of Toulouse producing firm, well-coloured wine.

🍇 **MIOUSAP**, Gascon rarity which has been rescued from oblivion but is still grown in extremely limited quantities. The wines can smell of peach brandy and have noticeable astringency.

🍇 **MISKET**, Bulgarian grape-scented variety that, despite its name, has no member of the MUSCAT family in its antecedents. It is a crossing of the native DIMIAT with RIESLING.

Red Misket, also used for perfumed white wines, is probably a pink-berried mutant of Misket and is relatively well established in Bulgaria with numerous local subvarieties such as **Sliven Misket** and **Varna Misket**. Misket is a speciality of the Sungurlare region.

🍇 **MISSION**, the original black grape variety planted for sacramental purposes by Franciscan missionaries in Mexico, the south west of the United States, and California in the 17th and 18th centuries. Mission was presumably of Spanish origin, imported to America by the conquistadores, and is important as a survivor from the earliest *vinifera* varieties to be cultivated in the Americas. It is identical to the PAÍS of Chile, is a darker-skinned version of the CRIOLLA CHICA of Argentina, and is thought by some to be the same as the MONICA of Spain and Sardinia. It was an important variety in California until the spread of phylloxera in the 1880s and there were still more than 1,000 acres/400 ha grown in the early 1990s, mainly in the south of the state and used principally for sweet wines. The wine made from Mission is not particularly distinguished but the variety has enormous historical significance.

🍇 **MOLETTE**, common variety in Savoie in the French alps, used particularly for the sparkling wines of Seyssel. The base wine produced is neutral and much improved by the addition of some ROUSSETTE.

🍇 **MOLINARA**, grown in the Veneto region of north east Italy, particularly for Valpolicella. Its wines tend to be high in acidity and it is only about a third as much planted as the more substantial CORVINA. RONDINELLA is the third Valpolicella grape.

🍇 **MOLL**, robust but potentially interesting Majorcan, also known as Prensal.

MONASTRELL, the main Spanish name for the black grape variety known in France as Mourvèdre and also as Mataro. See MOURVÈDRE for more details.

🍇 **MONDEUSE NOIRE**, one of the oldest and most distinctive grape varieties of Savoie, bringing an Italianate depth of colour and bite to the region in contrast to the softer reds produced by the Gamay imported only after phylloxera. The juicy, peppery wines are powerfully flavoured and coloured and are some of Savoie's few to respond well to careful small oak ageing (although when grown prolifically on Savoie's more fertile, lower sites Mondeuse can easily be a dull wine too, which may explain why the variety has been underrated). Some authorities claim that Mondeuse is identical to the REFOSCO of Friuli. The wines can certainly be extremely similar—and the extent of the House of Savoy in the 16th century would provide a historical explanation—but the theory is disputed. Most Mondeuse is characterful stuff sold as a varietal Vin de Savoie unusually capable of ageing. It may also be blended with Pinot Noir and Gamay in the wines of Bugey.

A **Mondeuse Blanche** can occasionally be found in Savoie and Bugey.

♀ MONICA, widely planted on Sardinia where some varietal Monica di Sardegna is thus labelled. The Italian vineyard census of 1990 found more than 6,000 ha/15,000 acres of Monica in 1990, but nearly 4,000 ha had been pulled out from 1982. The variety is thought to have originated in Spain (although it is not known in modern Spain) and some ampelographers think it may be identical to the historic MISSION of California. Its wines are undistinguished and should be drunk young.

♀ MONTEPULCIANO, vigorous vine planted over much of central Italy (31,000 ha/76,500 acres in 1990), recommended for 20 of Italy's 95 provinces but most widely planted in the Abruzzi, where it is responsible for the often excellent Montepulciano d'Abruzzo, and in the Marches, where it is a principal ingredient in such reds as Rosso Conero and Rosso Piceno. It is also grown in Molise and Apulia. The variety ripens too late to be planted much further north but can yield dependable quantities of well-priced, deep-coloured, well-ripened grapes with good levels of alcohol and extract (although some northern bottlers have a tendency to stretch it with less concentrated wine). It is sometimes called Cordisco, Morellone, Primaticcio, and Uva Abruzzi.

MONTONICO. See MANTONICO.

♀ MONTÙ, MONTUNI, indigenous to the plains of Emilia in north central Italy. Plantings totalled 1,200 ha/3,000 acres in 1990.

♀ MORAVIA, common grape in southern and central Spain planted on about 8,000 ha/20,000 acres. It produces rustic wines, particularly in south eastern La Mancha.

MORELLINO, local name for SANGIOVESE along the Tuscan coast.

MORELLONE, occasional name for Italy's MONTEPULCIANO.

♀ MORETO, undistinguished vine widely planted in Portugal, notably but not exclusively in Alentejo.

MORILLON, old name for PINOT NOIR in Champagne; old name for CHARDONNAY in much of France; common modern name for Chardonnay in Styria in southern Austria.

🍷 **MORIO-MUSKAT,** Germany's most popular MUSCAT-like vine variety by far, although it is quite unrelated to any true Muscat. Somehow Peter Morio's SILVANER × Weissburgunder (PINOT BLANC) crossing is almost overwhelmingly endowed with sickly grapiness that recalls some of Muscat's more obvious characteristics, even though its parents are two of the more aromatically restrained varieties. It was particularly popular with the eager blenders of the Pfalz and Rheinhessen in the late 1970s when its total German area reached 3,000 ha / 7,500 acres, and demand for Liebfraumilch was high. For a drop of Morio-Muskat in a neutral blend of MÜLLER-THURGAU and Silvaner can cheaply Germanize it. Total plantings fell to below 2,000 ha by 1990 and there are signs that this aggressively blowsy crossing may have had its day. If allowed to ripen fully, it can produce reasonably respectable varietal wines but it needs at least as good a site as Silvaner to achieve this. Must weights of Morio-Muskat are naturally low, although acidity is medium to high. The grapes can rot easily and ripen a week after Müller-Thurgau, which means that BACCHUS is a better alternative for Germany's cooler northern wine regions.

🍷 **MORISTEL,** light, loganberry-flavoured speciality of Somontano in the north of Spain. The vine is relatively frail and the wine produced oxidizes easily. May be most suitable as a fruity ingredient in blends.

MORRASTEL, main French synonym for Rioja's GRACIANO. It is also, confusingly, one of Spain's synonyms for MOURVÈDRE, although Monastrell is the more common Spanish name. Morrastel is the name used for the Graciano still grown in the Central Asian republic of Uzbekistan.

Morrastel-Bouschet, a much lesser crossing, is sometimes called simply Morrastel in southern France, where it was grown in considerable quantity in the mid 20th century, most notably in the Aude and Hérault *départements*. See Graciano for more details.

MORTÁGUA, western Portuguese synonym for CASTELÃO FRANCÊS, occasionally used for TOURIGA NACIONAL in Ribatejo.

🍷 **MOSCADELLO,** sometimes **MOSCADELLETO**, local strain of the MOSCATO BIANCO grape in and around Montalcino in central Italy. The firm Villa Banfi made an important investment in selling this sweet grapey white wine in the 1980s, and other producers of Brunello di Montalcino followed their lead. There is a fortified 'liquoroso' version.

MOSCATEL, Spanish Muscat, usually MUSCAT OF ALEXANDRIA. **Moscatel de Alejandría, Moscatel de España, Moscatel Gordo (Blanco), Moscatel de Málaga, Moscatel de Setúbal,** are all names for Muscat of Alexandria, although **Moscatel de Grano Menudo, Moscatel de Frontignan** are Spanish synonyms for MUSCAT BLANC À PETITS GRAINS.

🍷 **MOSCATEL DE AUSTRIA,** important variety in Chile, where it is the chief variety used for the local spirit, pisco. It is almost certainly the same as the TORRONTÉS Sanjuanino of Argentina and is valued for its productivity and the relative neutrality of its wine. Its thin-skinned grapes and compact bunches make it prone to rot in damper climates.

🍷 **MOSCATEL ROSADA,** quantitatively important and qualitatively very unimportant variety in Argentina, apparently unrelated to any known wine-making MUSCAT and much used for table grapes.

MOSCATO, Italian for MUSCAT.

🍷 **MOSCATO BIANCO,** sometimes called **Moscato di Canelli**, principal Italian name for the fine MUSCAT BLANC À PETITS GRAINS, the most planted Muscat in Italy, and the country's fourth most planted white grape variety with more than 13,000 ha/32,000 acres planted in 1990. Like MALVASIA, Moscato Bianco is ancient, versatile, and enjoys a geographical distribution that covers virtually the entire peninsula. Wines called Moscato are produced all over the country and are usually made from Moscato Bianco grapes. The light, refreshing, slightly sparkling Moscato d'Asti is one of its noblest incarnations. In the south and, especially, the islands, Italian Moscatos are typically golden and sweet.

Moscato Giallo and **Moscato Rosa** (both found in Alto Adige and often called Goldmuskateller and Rosenmuskateller respectively) are mutations with deeper-coloured berries.

MOSCATO DI ALEXANDRIA, Italian synonym for the lesser MUSCAT OF ALEXANDRIA, not widely planted but the grape of Moscato di Pantelleria.

🍷 **MOSCOPHILERO,** Greek vine with deep pink-skinned grapes used to make strongly perfumed white wine, notably on the high plateau of Mantinia in the Peloponnese, where conditions are sufficiently cool that harvest is often delayed until well into October. There are strong flavour similarities with fine MUSCAT but the origins of this distinct vine variety are as yet obscure. Small quantities of fruity light pink wine are also made from this spicy variety, which is also increasingly used as a blending ingredient in other parts of Greece.

MOSTER, occasional Austrian synonym for CHASSELAS.

🍷 **MOURISCO TINTO,** lesser port variety which produces red wines relatively light in colour in northern Portugal. The country's total plantings totalled nearly 5,200 ha/13,000 acres in 1992, many of them used for palish, tart table wines.

🍇 **MOURVÈDRE,** Spain's second most important black grape variety after Garnacha (Grenache) and once Provence's most important vine. The Spaniards call it Monastrell (and occasionally Morrastel or Morastell; although it has nothing to do with GRACIANO which is known as Morrastel in France). Mourvèdre is enjoying a resurgence of popularity, especially in southern France and, to a more limited extent, in California. In the New World it is often called Mataro.

The origins of this robust variety are almost certainly Spanish. Murviedro was a town near Valencia (Mataró is another near Barcelona). It is certainly easier to grow in Spain than in the cooler reaches of southern France for, although it is sensitive to low winter temperatures, it buds and ripens extremely late, a week later even than Carignan according to Galet. Provided the climate is warm, the upright, vigorous Monastrell adapts well to a wide range of soils and recovers well from spring frost. It is susceptible to both downy and powdery mildews, which is much less of a problem in hot Spanish vineyards than in much of France.

The wine produced from Monastrell's small, sweet, thick-skinned berries tends to be heady stuff, high in alcohol, tannins, and flavour when young and well capable of ageing provided oxidation is carefully avoided in the winery. The wine can have a dangerously gamey, not to say animal, note. As Monastrell, it is planted on well over 100,000 ha/250,000 acres in Spain, especially in the Murcia, Alicante, Albacete, and Valencia regions and all over the Levante. It is the principal black grape variety in such DO wines as Alicante, Almansa, Jumilla, Valencia, and Yecla.

Mourvèdre needs France's warmest summers to ripen fully. It dominated Provence until the arrival of phylloxera and the search for productive vines to supply the burgeoning market for cheap table wine. For many decades it marked time in its French enclave Bandol but is now regarded as an extremely modish and desirable 'improving variety' throughout the Languedoc-Roussillon, especially now that clones have been selected that no longer display the inconveniently variable yields that once resulted from degenerated vine stock. Between 1968 and 1988 total French plantings increased from 900 to 5,600 ha, spread between Provence, the southern Rhône, the Languedoc, and Roussillon.

In southern France Mourvèdre produces wines considered useful for their structure, intense fruit and, in good years, perfume often redolent of blackberries. The structure in particular can be a useful foil for Grenache in Provence and Cinsaut further west. In Bandol it is typically blended with both of these, and the statutory minimum for Mourvèdre is now 50 per cent. Mourvèdre is condoned in a host of appellation contrôlée regulations all over the south of France from Collioure to Coteaux du Tricastin. It is at its most successful playing a supporting role, being fleshier than Syrah, tauter than Grenache and Cinsaut, and infinitely more charming than Carignan, although varietal Mourvèdres are by no means unknown.

For many years Australia's Mataro (also known as Esparte) was treated with some disdain, and only about 600 ha/1,500 acres were yet to be ripped out in 1990. It is slowly gaining stature, following the example of Shiraz, by virtue of its Rhôney image. This has already happened in California. Although grown at least since the 1870s, California's unfashionable Mataro was fast disappearing until the 'Rhône Rangers' made the connection with Mourvèdre and pushed up demand for wine from these historic stumps, notably in Contra Costa County immediately east of San Francisco. There were also considerable new plantings in the early 1990s thanks to demand from the likes of Bonny Doon and Cline Cellars. By 1992 the state's total plantings were nearly 300 acres/120 ha.

Galet notes that there may be some Mourvèdre in Azerbaijan, although that is not what it is called.

℣ MOUYSSAGUÈS, ancient vine of the Aveyron in the harsh uplands of South West France. Its wine can be dark and tough and it has been largely abandoned because it does not graft well.

MOZA FRESCA, Valdeorras name for DOÑA BLANCA.

℣ MÜLLER-THURGAU, decidedly mediocre but gruesomely popular German crossing developed in 1882 for entirely expedient reasons by a Dr Hermann Müller, born in the Swiss canton of Thurgau but then working at the German viticultural station at Geisenheim. His understandable aim was to combine the quality of the great RIESLING grape with the viticultural reliability, particularly the early ripening, of the SILVANER. Most of the variety's synonyms (Rivaner in Luxembourg and Slovenia, Riesling-Sylvaner in New Zealand and Switzerland, Rizlingszilvani in Hungary) reflect this combination. Since then some authorities have argued that he actually crossed two strains of Riesling rather than, as he thought, Riesling with Silvaner, but whatever the ingredients the recipe resulted in a variety all too short on Riesling characteristics (indeed much shorter on such elegant raciness than the more recent crossings EHRENFELSER, FABER, KERNER, and well-ripened SCHEUREBE).

The vine certainly ripens early, even earlier than Silvaner. Unlike Riesling it can be grown anywhere, producing prodigious quantities (sometimes double Riesling's common yield range of 80 to 110 hl/ha (4.5–6 tons/acres)) of extremely dull, flabby wine. Müller-Thurgau usually has some vaguely aromatic quality, but the aroma can often be unattractively mousey in Germany's high-yielding vineyards and is more reliably clean and pure in the variety's other spheres of influence New Zealand, Alto Adige, and west Washington State, where growers are less demanding in terms of quantity.

Müller-Thurgau was not embraced by Germany's growers until after the Second World War when the need to rebuild the industry fast presumably gave this productive, easily grown vine allure. In the early 1970s it even overtook the great Riesling in total area planted (having for some time produced far more wine in total) and remained in that position throughout the 1980s, although by the end of that decade there were already signs of disaffection with the grape on which German wine industry was commercially based. Occasionally a German Müller-Thurgau could be said to express something—usually something territorial rather than anything inherent in the grape—but this bland vehicle for quantity above quality was substantially responsible for the decline in Germany's reputation as a wine producer in the 1970s and 1980s. Typically blended with a little of a more aromatic variety such as MORIO-MUSKAT and sweetened with Süssreserve, or grape juice, Müller-Thurgau was transformed into oceans of sugarwater. In 1990 it still occupied nearly a quarter of Germany's vineyards, a third of all Baden plantings, nearly half (and still increasing) of those in Franken, but also nearly a quarter of the Mosel-Saar-Ruwer and Nahe regions.

The wood is much softer than Riesling's and can easily be damaged by hard winters. The grapes rot easily (as can be tasted in a number of examples from less successful years) and the vine is susceptible to downy mildew, black rot, and, its own bane, Roter Brenner, but it will presumably continue to flourish while there is a market for cheap German wine.

Outside Germany it can taste quite palatable, if rarely exciting. A handful of Italians manage it in the Alto Adige, where high altitudes keep the grapes on the vine for long enough for them to retain acidity while developing some perceptible fruit flavours. It is also increasingly planted in Friuli and is grown as far south as Emilia-Romagna. This foreign-sounding variety, sometimes called Riesling-Sylvaner, to the Germans' horror, has its followers among Italy's fashion-conscious connoisseurs.

The variety thrives all over central and eastern Europe. It is planted, appropriately enough, in Switzerland, playing an increasingly important role in the vineyards of the German-speaking area in the north and east. Only the native Grüner Veltliner is more important in Austria, where it still comprises nearly one vine in every ten but is rarely responsible for wines of much intrinsic interest. Across Austria's southern border, it is also grown in Slovenia and is even more important to the east and north of Austria in Slovakia and, particularly, Hungary, which is probably the world's second most important grower of this uninspiring grape. As Rizlingszilvani, it covers thousands of ha of vineyard around Lake Balaton and produces lakesful of flabby Badacsonyi Rizlingszilvani.

Müller-Thurgau was planted enthusiastically by New Zealand grape growers on the recommendation of visiting German experts as a preferable susbstitution for the hybrids that were all too prevalent in the country's nascent wine industry of the 1950s and 1960s. It was the country's dominant grape variety until 1993 when Chardonnay overtook it in area planted, as it is a far more valuable crop. It would be difficult to argue that New Zealand's 'Riesling-Sylvaner' is ever a very complex wine but it does usually display a freshness lacking in German examples, despite its customary similar reliance on sweet grape juice.

Elsewhere in the New World, most growers are not driven by the need for early ripening varieties (and would find the flab in the resultant wine a distinct disadvantage) although some Oregon growers have experimented successfully with it, and fine examples have been produced in the western Puget Sound vineyards of Washington State.

Northern Europe's two smallest and coolest wine producers, England and Luxembourg, depend heavily on the reliably early ripening Müller-Thurgau, which (called Rivaner in Luxembourg) is the most planted variety in each country. Examples made this far from the equator benefit from their additional acidity.

MÜLLERREBE, which translates from German as 'miller's grape', is the common, and logical, name for Germany's increasingly planted selection of Pinot MEUNIER. (Schwarzriesling is another German synonym.) It is most common in Württemberg, which has its own low-yielding mutation called Samtrot, literally 'red velvet', of which nearly 100 ha/250 acres were planted in 1990.

MUSCADEL or **MUSKADEL,** occasional South African name for MUSCAT BLANC À PETITS GRAINS of any colour.

🍷 **MUSCADELLE,** famous also-ran third grape responsible, with SÉMILLON and SAUVIGNON BLANC, for the sweet white (and duller dry white) wines of Bordeaux and Bergerac. Like all of Bordeaux's white grape varieties except for Sauvignon, its star is waning, but not nearly so rapidly as that of UGNI BLANC or COLOMBARD. Four out of every five Muscadelle vines grown in Bordeaux are not in the great sweet white wine area of Sauternes and Barsac, but in the unfashionable and vast Entre-Deux-Mers, including such lesser sweet white appellations as Premières Côtes de Bordeaux, Cadillac, Loupiac, and Ste-Croix-du-Mont. Muscadelle is also being pulled up at quite a rate in the Dordogne, including such appellations as Monbazillac, but it is still relatively more important to Bergerac than to Bordeaux.

The variety, unrelated to any member of the MUSCAT family, shares a vaguely grapey aroma with them but has its origins in Bordeaux. The usefully productive Muscadelle leafs late and ripens early and has never demonstrated great subtlety in the wines it produces. Indeed its use is almost exclusively in blends, adding the same sort of youthful fruitiness to south western sweet whites as MEUNIER does to the north eastern sparkling whites called champagne.

The variety is grown widely but not importantly in eastern Europe, but in only one obscure corner of the wine world does

Muscadelle produce sensational varietal wine, the strong, sweet, dark, barrel-aged Liqueur Tokays of Australia. For years Australians thought the grape they called Tokay was the Hungarian HÁRSLEVELŰ, but the French ampelographer Paul Truel identified it as Muscadelle in 1976. There were 400 ha / 1,000 acres of Muscadelle in Australia in 1990, almost exactly the same area as was planted with the BROWN MUSCAT from which Liqueur Muscat is made. Much of this Muscadelle is grown in South Australia but there are plantings in north east Victoria, famous for these Australian stickies, too.

On the same vine classification mission to Australia, Truel identified vines imported from California as 'Sauvignon Vert' as Muscadelle, and it is probable that California's minuscule plantings of the variety known there as Sauvignon Vert (a total of less than 100 acres / 40 ha in 1991) are in fact of the third Bordeaux variety rather than Sauvignon Vert (or TOCAI Friulano).

♛ MUSCARDIN, minor southern Rhône variety officially but usually theoretically allowed in to Châteauneuf-du-Pape, making a pale but scented contribution to the blend at Château de Beaucastel. Galet notes a similarity with MONDEUSE.

MUSCAT, one of the world's great and historic names, both of grapes and wines. Indeed Muscat grapes—and there are at least four principal varieties of Muscat, in several hues of berry—are some of the very few which produce wines that actually taste of grapes. MUSCAT HAMBURG and MUSCAT OF ALEXANDRIA are raised both as wine grapes and table grapes (although it has to be said that Hamburg is much better in the second role). MUSCAT BLANC À PETITS GRAINS is the oldest and finest, producing wines of the greatest intensity, while MUSCAT OTTONEL, paler in every way, is a relative parvenu.

Muscat grapes were probably the first to be distinguished and identified and have grown around the Mediterranean for many, many centuries. With such strongly perfumed grapes, described in French as *musqué* as though they were actually impregnated with musk, Muscat grapes have always been attractive to bees and it was almost certainly Muscat grapes that the Greeks described as *anathelicon moschaton*, and Pliny the Elder as *uva apiana*, 'grape of the bees'. Some even theorize that Muscat derives its name from *musca*, the Latin for flies, which are also attracted to these scented grapes.

Muscat wines, carrying many different labels including Moscato (in Italy) and Moscatel (in Iberia), can vary from the refreshingly low-alcohol, sweet, and frothy Asti Spumante, through Muscat d'Alsace and its fashionable bone dry mimics made from a Muscat surplus resulting from the world's worship of the light and dry, to sweet wines with alcohol levels between 15 and 20 per cent. Since a high proportion of the world's Muscat is dark-berried, and since a wide variety of wood ageing techniques are used, such wines can vary in colour from palest gold (as in some of the more determinedly modern Muscats de Frontignan) to deepest brown (as in some of Australia's Liqueur Muscats).

Most Muscat vines need relatively hot climates (although see MUSCAT OTTONEL) and there either are or have been many famous Muscats around the Mediterranean.

🍇 **MUSCAT BLANC, MUSCAT BLANC À PETITS GRAINS, MUSCAT BLANC À PETITS GRAINS RONDS**, is the oldest and noblest variety of Muscat with the greatest concentration of fine grape flavour, hinting at orange-flowers and spice. Its berries are, as its name suggests, particularly small, and round as opposed to the oval berries of MUSCAT OF ALEXANDRIA. But its berries are not, as its principal ampelographical name suggests, invariably white. In fact there are pink-, red- and black-berried versions (although the dark berries are not so deeply pigmented that they can produce a proper red wine) and some vines produce berries whose colour varies considerably from vintage to vintage. Many synonyms for the variety include reference to the yellow or golden (*gallego, giallo, gelber*) colour of its berries. And Brown Muscat is one of Australia's names for a Muscat population that is more dark than light and resembles South Africa's MUSKADEL in that respect (thereby providing more evidence of early viticultural links between these two southern hemisphere producers). Other names for the variety in its many different habitats include **Muscat of Frontignan**, Frontignac, **Muscat Lunel, Muscat Blanc, Muscat d'Alsace, Muscat Canelli, Muskateller**, Moscato Bianco, Moscato d'Asti, Moscato di Canelli, Moscatel de Grano Menudo, Moscatel de Frontignan, Moscatel Branco, White Muscat, and Muscadel or Muskadel (in South Africa). Any Muscat with the words Alexandria, Gordo, Romain, Hamburg, or Ottonel in its name is *not* this superior variety.

This particular Muscat is almost certainly the oldest known wine grape variety, and the oldest cultivated in France, having been established in Gaul around Narbonne, notably at Frontignan, by the Romans—and possibly even before then brought to the Marseilles region by the Greeks. Muscat Blanc has clearly been established for many centuries round the Mediterranean, where its early budding poses few problems. It was certainly already widely esteemed in the vineyards of Roussillon by the 14th century, and dominated them until the 19th (apparently predating the arrival of MALVASIA from the east). It is Piedmont's oldest documented variety. That it is recorded as growing in Germany, as Muskateller, as early as the 12th century, and is the first documented variety grown in Alsace, in the 16th century, suggests that spring frosts may have been less common then for the variety has now been replaced by the most accommodating Muscat Ottonel in Alsace and has all but disappeared from a Germany apparently in thrall to the flashily ersatz Morio-Muskat crossing (there were just 50 ha/120 acres of 'Gelber Muskateller' by 1990).

Muscat Blanc also yields more conservatively than other

Muscats and is sensitive to a wide range of diseases, which has naturally limited its cultivation. As Moscatel de Grano Menudo it is still grown in Spain but to a limited extent. Most Spanish wines labelled Moscatel are made from MUSCAT OF ALEXANDRIA.

Muscats of various sorts are grown widely in the ex-Soviet Union, particularly MUSCAT OTTONEL and the variety decribed as **Muscat Rose,** the pink-skinned form of Muscat Blanc. Muscat of some sort is grown in Russia (as Tamyanka), Ukraine, Moldova, Kazakhstan, Uzbekistan, Tajikistan, and Turkmenistan. Early 20th century bottles labelled White Muscat and Pink Muscat from the Massandra winery in the Crimea show some of the dramatic ageing potential of this variety. Muscat Blanc is also grown, as Tămîioasă Alba, in Romania, where Ottonel predominates, and in parts of what was Yugoslavia as Zutimuscat and Beli Muscat. In the heart of Hapsburg country, Muscat Ottonel has held sway until recently. In the early 1980s Austrians realized the greater potential inherent in their small plantings of Muskateller. Recently, dry, racy Muskatellers from Styria and occasionally the Wachau have been some of the country's most sought after and plantings are increasing. In Hungary too Ottonel, simply known as Muskotaly, dominates except in the Tokaj district where Muscat Blanc, there called Lunel or Sargamuskotaly (Yellow Muscat) is grown on about 200 ha / 500 acres. Some varietal Muscats of sweet Aszú quality are made from the few hundred ha that supplement the Furmint and Hárslevelű that are the main ingredients in this extraordinary wine.

If anywhere could be said to be Muscat's homeland it is Greece, however, and here, although it is today grown alongside Muscat of Alexandria (which is the prime Cypriot Muscat), Muscat Blanc à Petits Grains is accorded the honour of being the only variety allowed in Greece's most rigidly controlled Muscats such as those of Samos, Patras, and Cephalonia. For the moment Greek Muscat, like its many variations on the MALVASIA theme, is almost invariably sweet, alcoholic, and redolent of history, but drier versions more suited to drinking with food are expected.

This is the Muscat that, as MOSCATO, predominates in Italy, which grows about 30,000 ha / 75,000 acres of it, most profitably as underpinning for the sweet sparkling wine industry. The light, frothy Asti Spumante, the subtler Moscato d'Asti and other *spumante* and *frizzante* all over north western Italy demonstrate another facet of the variety's character. Various forms of MOSCATO can be found throughout Italy but most of its produce in the south and inland belongs to the richer, Mediterranean school of wines.

This school represents the traditional face of Muscat Blanc in France but, contrary to almost all other white grape varieties, this Muscat has been gaining ground, chiefly because of the development of less traditional forms of Muscat wine. France's total plantings of Muscat Blanc have been steadily increasing, thanks to the development of a virtually CLAIRETTE-free grapey version of the Rhône's fizzy Clairette de Die.

Muscat Blanc has also been supplanting the still more widely planted Muscat of Alexandria in Roussillon where it is the superior Muscat ingredient in the many strong sweet wines, particularly Muscat de Rivesaltes. In the Languedoc and southern Rhône too its increasing area of vineyard reflects increased demand for the golden sweet Muscats of Beaumes-de-Venise, Frontignan, Lunel, Mireval, and St-Jean-de-Minervois, in which it is the exclusive ingredient. But Muscat has been enjoying a new lease of life in the Midi vinified dry, without the addition of grape spirit to preserve its natural sweetness and add extra alcohol. Such wines with an easily recognizable aroma all too rare in southern French whites, together with their fashionably dry, light impact on the palate, have provided popular inexpensive alternatives to the dry Muscat (usually Ottonel) of Alsace.

In the New World the variety grows, as Brown Muscat and Frontignac with all hues of grape skins, in Australia where it is capable of the great Liqueur Muscats as well as in South Africa, where the wines it produces are known as both Frontignac and Muscadel or Muskadel.

The 1,000 acres / 400 ha of Muscat used for California wine production are mainly in the Central Valley, and almost all of this superior Muscat variety. It was once variously called Muscat Frontignan and Muscat Canelli but has now been officially separated into Muscat Blanc and a little Orange Muscat, the Quady winery having based its fortunes on making a sweet wine from the latter. Madera County in the Central Valley is an important source of richer Muscat Blanc, while Paso Robles has also proved a congenial home.

🍷 MUSCAT HAMBURG, is the lowest of the wine-producing Muscats. It comes exclusively in black-berried form and is far more common as a table grape than a wine grape. Its chief attribute is the consistency of its plump and shiny dark blue grapes, which can well withstand long journeys to reach consumers who like black Muscat-flavoured grapes. In France it is the second most important table grape after Chasselas and it is also relatively important as a table grape in Greece, in eastern Europe, and Australia. It was extremely popular as a greenhouse grape in Victorian England, where it occasionally took the name of Snow or Venn, two of its more successful propagators.

In the world of wine production its importance is limited but it does produce a fair quantity of light, grapey red throughout eastern Europe. In China, crossed with the indigenous *Vitis amurensis*, it has spawned generations of varieties adapted for wine production.

🍷 MUSCAT OF ALEXANDRIA, is a Muscat almost as ancient as MUSCAT BLANC À PETITS GRAINS but its wine is generally inferior. In hot climates it can thrive and produce a good yield of extremely ripe grapes but their chief attribute is sweetness rather than subtlety of flavour. (In cooler climates its output can be seriously affected by coulure, millerandage, and a range of fungal diseases.) Wines made from this sort of Muscat tend to be strong, sweet, and unsubtle. The aroma is vaguely grapey but can have slightly feline overtones of geranium rather than the more lingering bouquet of Muscat Blanc: marmalade as opposed to orange blossom.

Some indication of its lack of finesse as a wine producer is the fact that a considerable proportion of the Muscat of Alexandria grown today is destined for uses other than wine. California for example uses its 5,000 acres/2,000 ha for raisins. Chile distils most of its Muscat of Alexandria to make pisco, the national spirit. The variety is even grown under glass in climates as inimical as the Dutch and British to provide grapes for the fruit bowl.

As its name suggests, Muscat of Alexandria is thought to have originated in Egypt and was disseminated around the Mediterranean by the Romans, hence its common synonym **Muscat Romain**. Its southern Italian synonym Zibibbo echoes the North African Cape Zibibb. Today it is most important to wine industries in that old arc of maritime history Iberia, South Africa, and Australia, where its chief respective names are Moscatel, Hanepoot, and **Muscat Gordo Blanco** or Lexia, a particularly Australian contraction of the word Alexandria. Spain has a considerable area planted with the variety, but only about half of this serves the wine industry, typically with sweet Moscatels of various sticky sorts. Muscat of Alexandria's various Spanish synonyms include Moscatel de España, Moscatel Gordo (Blanco), and most importantly Moscatel de Màlaga (in which shrinking wine zone Muscat produces probably its finest Spanish wine).

In Portugal its most famous incarnation is Moscatel de Setúbal, but Portugal's Muscat of Alexandria grapes have also been harnessed to produce aromatic, dry, much lower-alcohol Muscats whose prototype João Pires was developed, significantly, by an Australian wine-maker who knew well how to transform one of Australia's most planted grape varieties into an early-picked, crisp, technically perfect table wine. This is the fate of the majority of Australia's more than 3,500 ha/8,600 acres of Gordo Blanco, once used mainly for fortified wines, although from cooler vineyards it can produce sound, unfortified wines that are sweet because late-picked too. With SULTANA, Muscat Gordo Blanco is a mainstay of Australia's hot, irrigated vineyards and the proportion vinified as opposed to dried, varies with the annual ebb and flow of market forces. A much higher proportion, almost always the majority, of Muscat Gordo Blanco is made into wine than of Sultana. The wine produced is typically used for bulking out more glamorous grape varieties.

Muscat of Alexandria, often called Hanepoot, is the dominant Muscat in South Africa and, although it is losing ground, it was still the country's sixth most planted variety in the early 1990s covering well over twice as much ground as Cabernet Sauvignon, for example. For years it provided sticky, raisiny wines for fortification, as well as everything from grape syrup to raisins. Today some drier, lighter wines are also made from it.

Moscatel de Alejandría was much more important in Chile before the variety there called MOSCATEL DE AUSTRIA supplanted it as chief ingredient in pisco. Total plantings were well below 2,000 ha by the early 1990s. It is also grown to a relatively limited extent in Argentina (where MOSCATEL ROSADA is the dominant Muscat), Peru, Colombia, Ecuador, and even Japan.

Although Muscat Blanc is more important in Greece, Muscat of Alexandria is grown widely there. It is also the Muscat that predominates in Turkey, Israel, and Tunisia, although in much of the Near East nowadays these grapes are eaten rather than drunk. The rich, dark Moscato di Pantelleria is geographically closer to Tunisia than Sicily which administers it and is made from Muscat of Alexandria, or Zibibbo as it is known in much of (southern) Italy. Italy as a whole probably grows about a third as much of this lesser Muscat than the true Moscato Bianco that predominates in the north.

In France total plantings of **Muscat d'Alexandrie**, or Muscat Romain, have remained at about the same level as Italy's, just over 3,000 ha / 7,400 acres almost exclusively in Roussillon, since the 1960s. Although Muscat Blanc is catching up, Muscat of Alexandria is still the dominant Muscat in this most Spanish corner of France where at one time its grapes were left to raisin on the vine before adding their distinctive flavour to the highly prized local wines. It is most obvious in Muscat de Rivesaltes but is also blended with other varieties, chiefly GRENACHE of all hues, to produce such strong sweet wines as Banyuls, Rivesaltes, and Maury. It was the stagnation of sales of such wines in the late 1970s that provided a catalyst for today's southern French dry Muscats (see also MUSCAT BLANC À PETITS GRAINS).

🍇 **MUSCAT OTTONEL**, is the palest of all the Muscats both in terms of the colour of wine produced and in terms of its character. Its aroma is altogether more vapid than the powerful grapey perfumes associated with MUSCAT BLANC À PETITS GRAINS and MUSCAT OF ALEXANDRIA. It was bred as recently as 1852 in the Loire, probably as a table grape from CHASSELAS and the distinctly ordinary Muscat de Saumur, according to Galet.

Its tendency to ripen earlier than these other two Muscats has made it much easier to cultivate in cooler climates and nowadays Muscat Ottonel is virtually the only Muscat cultivated in Alsace. This low-vigour vine, which does best in deep, damp soils, is also grown in eastern Europe, notably in Austria where there are still substantial plantings. Until the 1980s it was revered to the exclusion of true Muscat Blanc, or Muskateller, particularly for its sweet wines made from nobly rotten grapes grown in the Neusiedlersee region which can be very fine. It may well be that it is at its best as a late harvest wine for there are some superior, apparently long-living examples from both Hungary and Romania

where the variety is often known, respectively, as Muskotaly and
TĂMÎIOASĂ Ottonel. Hungary had about 3,300 ha/8,100 acres of
Ottonel and Romania almost 6,000 ha/15,000 acres of Muscat
Ottonel in 1993. In Alsace, however, Vendange Tardive (Late
Harvest) Muscat tends to remain a theoretical possibility. One of
the most widely planted Muscats in the former Soviet Union is
Ottonel, often known as Hungarian Muscat. It is planted in Russia,
Ukraine, Moldova, Kazakhstan, Uzbekistan, Tajikistan, and
Turkmenistan.

MUSKADEL, South African name for MUSCAT BLANC À PETITS
GRAINS, often a dark-berried form.

MUSKATELLER, German for MUSCAT, almost invariably the
superior MUSCAT BLANC À PETITS GRAINS, or some mutation of it.
Gelber Muskateller, for example, is the gold-skinned version
which is increasingly recognized as superior to MUSCAT OTTONEL
in Austria where it is particularly popular in Styria. In Germany,
homeland of MORIO-MUSKAT, Gelber Muskateller is a distinctly
minority interest, and there is ever less of the red-skinned **Roter
Muskateller**.

MUSKAT-OTTONEL is what Germans call their minuscule
plantings of MUSCAT OTTONEL.

MUSKAT-SILVANER or **MUSKAT-SYLVANER** is, tellingly, the
common German language synonym for SAUVIGNON BLANC and is
grown in Germany and Austria to a very limited but increasing
extent, notably in Styria where it is more likely to be called
Sauvignon Blanc.

MUSKOTÁLY, Hungarian name for MUSCAT, usually MUSCAT
OTTONEL but also occasionally for a yellow-berried form of MUSCAT
BLANC À PETITS GRAINS, here called Muscat Lunel.

MUSQUÉ is a French term meaning both perfumed, as in musky,
and MUSCAT-like. Many vine varieties, including CHARDONNAY, have
a Musqué mutation which is particularly aromatic and may add to
the variety's own characteristics a grapey, heady scent.

N

NAGYBURGUNDI, Hungarian name for BLAUFRÄNKISCH.

NAPA GAMAY. See GAMAY.

⚘ **NASCO,** ancient Sardinian vine making soft wines around
Cagliari.

⚘ **NEBBIOLO,** great, geographically sensitive vine responsible
for some of the finest and longest-lived wines in the world. The

wines are very deep coloured, high in tannins and acidity in youth, but can evolve after years in bottle into some of the most seductively scented wines in the world, with a bouquet ranging from tar through violets to roses. Nebbiolo is native, and almost confined, to Piedmont in north west Italy, where it is the undisputed king in a kingdom of distinctive vine varieties. It was recorded as a celebrated vine of the region as early as the 14th century and probably takes its name from *nebbia*, or fog, a frequent phenomenon in Piedmont in October when the grape is harvested. Nebbiolo is of prime qualitative, but almost negligible quantitative, importance to Italy's wine industry.

Even within Piedmont, Nebbiolo is restricted to a few selected areas, and the variety rarely accounts for more than 3 per cent of Piedmont's total wine production, a mere fraction of the amount of BARBERA produced, for example. Nebbiolo is always a late ripener, with harvests that regularly last well past the middle of October, and the variety is accordingly granted the most favourable hillside exposures, facing south to south west. Perhaps as important as the vineyard site, however, are the soils in which the variety is planted: Nebbiolo has shown itself to be extremely fussy and has given best results only in the calcareous marls to the north and south of Alba on the right bank of the Tanaro in Barbaresco and Barolo respectively. Here Nebbiolo-based wines reach their maximum aromatic complexity, and express a fullness of flavour which balances the relatively high acidity and substantial tannins which are invariably present. Wines made in the greater Nebbiolo delle Langhe zone or even the more specific Nebbiolo d'Alba are much less intense and long-lasting than the great and highly varied wines from specific vineyards within Barolo and Barbaresco. And when planted on the sandier soils of the Roero district on the left bank of the Tanaro, Nebbiolo is particularly light and soft. (There are obvious parallels between Nebbiolo in Piedmont and PINOT NOIR in Burgundy in terms of the vines' sensitivity to site.)

Good Nebbiolo wines are also produced in varying soil types in the hills on the left and right banks of the Sesia river to produce such wines as Boca, Bramaterra, Fara, Gattinara, Ghemme, Lessona, and Sizzano. Here Nebbiolo is called Spanna and is usually blended with softer VESPOLINA and/or BONARDA grapes.

Nebbiolo, often called Picutener, also plays the leading role in the tiny Carema zone on the border of the Valle d'Aosta, and in its equally minuscule neighbour Donnaz. In the far north of Lombardy in the Valtellina (where it is known as Chiavennasca) is the only sizeable zone where Nebbiolo is cultivated outside Piedmont.

Nebbiolo is hardly known elsewhere in Italy, although it is an ingredient in Lombardy's Franciacorta cocktail, and the innovative Veneto wine-maker Giuseppe Quintarelli makes a Recioto from dried Nebbiolo grapes.

Three principal clones of Nebbiolo are conventionally identified (although there is still much work to be done in this respect): Lampia, Michet, and the fast-disappearing Rosé, whose wines are inconveniently pale. Michet is a form of virus-affected Lampia which, while producing small crops and particularly intense aromas and flavours, does not adapt itself to all soils. Most producers prefer to rely on a careful mass selection in their vineyards rather than staking their future on a single clone.

The total area planted with Nebbiolo declined in the 1980s, to about 5,200 ha/13,000 acres in 1990—about half the area planted with Piedmont's DOLCETTO, and about a tenth the total Italian area planted with Barbera.

The quality of Barolo and Barbaresco has inspired vine-growers all over the world to experiment with Nebbiolo, but the resulting wines have so far lacked most of the variety's best qualities on home ground. Nebbiolo has somewhat reluctantly accompanied Barbera to both North and South America. Few California examples have so far demonstrated much of the grape's intrinsic worth, although the fashion for all things Italian has provided a strong incentive to rectify this situation. In South America, high yields have tended to subsume the variety's quality. The few hundred ha planted in Argentina are mainly in San Juan province.

 NEGOSKA, Greek variety that is a softening ingredient, with XYNOMAVRO in the wines of Goumenissa. Very fruity, alcoholic wines.

NEGRA DE MADRID, synonym for GRENACHE, or Garnacha, in the Madrid region.

NEGRA MOLE and NEGRAMOLL, undistinguished Iberian dark-skinned grape variety. See TINTA NEGRA MOLE.

 NEGRARA, declining north east Italian speciality, of which **Negrara Trentina** is the most common.

 NÉGRETTE, distinctive speciality of the vineyards north of Toulouse in South West France. It dominates the blend (with Bordeaux varieties) in Côtes du Frontonnais and in the wines of Lavilledieu it must constitute at least 35 per cent. Wine made from Négrette is more supple, perfumed, and flirtatious than the more famous south western black grape variety Tannat, and is probably best drunk young, with its fruit, sometimes described as having a slightly animal flavour, unsuppressed by heavy oak ageing. As a vine, the variety is inconveniently prone to powdery mildew and rot and is therefore better suited to the climate of Toulouse than to damper regions. The variety sold as Pinot St George in California in the 1960s and 1970s was thought to be none other than Négrette by French ampelographer Galet on his 1980 tour of American vineyards.

NEGROAMARO

🍷 **NEGROAMARO**, or **NEGRO AMARO**, useful southern Italian vine that is the country's sixth most planted with a total of 31,000 ha/76,500 acres in 1990. This is *the* variety of Apulia (although PRIMITIVO is also important) and if carefully vinified, can produce wines of great intensity—and alcohol. Although it has traditionally been used for blending, it can produce vigorous red wines worthy of ageing as well as some lively rosé, mainly from the heel of Italy.

🍾 **NEHERLESCHOL**, extremely ancient Middle Eastern grape planted experimentally at Mas de Daumas Gassac in the Languedoc. Has enormous bunches.

🍷 **NERELLO**, important and productive Sicilian variety. **Nerello Mascalese** is more widely planted than **Nerello Cappuccio** and is concentrated in the north east of the island. The wines produced tend to lack the concentration of NERO D'AVOLA, although they are usually high in alcohol. Most of the wine is used for blending. Total plantings of the two varieties together were about 18,000 ha/44,500 acres in 1990.

🍷 **NERO D'AVOLA**, one of the best red wine grapes of Sicily, also known as Calabrese, suggesting origins in Calabria on the mainland. Total plantings of the variety fell by a third in the 1980s to about 14,000 ha/34,500 acres in 1990 (only a fraction of Sicily's vineyard devoted to the principal white grape CATARRATTO), but quality-minded producers on the island value the body and ageing potential which Nero d'Avola can bring to a blend. Varietal Nero d'Avola has shown itelf a fine candidate for barrel maturation, with fine aromas and real potential for ageing.

🍾 **NEUBURGER**, sometimes distinguished variety grown almost exclusively in Austria, where it was the fifth most common white wine grape in 1993. Quite possibly an accidental crossing, of Weissburgunder (PINOT BLANC) × SILVANER which makes wine that tastes like an even fuller bodied Weissburgunder. It ripens relatively early and more decisively than GRÜNER VELTLINER, Austria's most popular vine. It is grown in most of Austria's wine districts other than Styria.

🍷 **NEYRET**, rare vine occasionally found in the south east of the Valle d'Aosta.

🍾 **NIAGARA**, American hybrid of CONCORD × Cassady, a *vinifera* variety created in Niagara, New York in 1872. Today it is the most successful native white variety in New York State even though the wines produced are marked by a very foxy flavour. The vine is vigorous, productive, and withstands low temperatures well, although not as well as Concord. Niagara is the most planted white wine grape in Brazil.

🍷 **NIELLUCCIO**, Corsica's third most planted dark-berried vine, probably brought there from the Italian mainland, presumably by the Genoese, who ruled the island until the late 18th century, as it is ampelographically identical to the SANGIOVESE of Tuscany. It represented only 14 per cent of all Corsican vines in 1988, however, thanks to the domination of CINSAUT and CARIGNAN imported by French immigrants from North Africa in the 1960s and 1970s. Often blended with the, arguably more interesting, other major indigenous red wine variety SCIACARELLO, it constitutes an increasing proportion of the island's appellation contrôlée reds and, particularly, rosés, for which it is especially suitable. It is the principal ingredient in Patrimonio, on whose clay-limestone soils it thrives. It buds early and ripens late and is therefore susceptible to late frosts in spring and rot during the harvest.

♀ **NINCUSA,** minor grape grown on the Dalmatian coast of former Yugoslavia.

NOBLE, occasional synonym for PINOT.

♀ **NOBLING,** 1939 German crossing of SILVANER × Gutedel (CHASSELAS) that is declining in importance even in Baden, where there were more than 100 ha/250 acres in the late 1970s. It can ripen relatively well and yet retain acidity but it needs a relatively good site that can be used more profitably for more fashionable or productive varieties.

NOIR, French for black and therefore a common suffix for dark-berried grape variety names.

NOIRIEN is, most commonly, the name given to the PINOT family of grape varieties found primarily in eastern France that are related to or closely associated with PINOT NOIR: PINOT GRIS, PINOT BLANC, AUXERROIS, and (although it is not related) CHARDONNAY. It is also a synonym for Pinot Noir and, more misleadingly, **Noirien Blanc** is used as a synonym for Chardonnay.

♀ **NOSIOLA,** speciality of the Trentino region in northern Italy responsible for the wine of the same name, as well as small amounts of Sorni Bianco. The wines have more aroma than body, and finish with slight bitterness.

♀ **NURAGUS,** Sardinian speciality grown principally to produce the unremarkable varietal Nuragus di Cagliari. Total plantings halved during the 1980s to a total of about 8,700 ha/21,500 acres in 1990.

O

🍇 **OEILLADE,** occasional synonym for CINSAUT, especially when sold as a table grape and also an almost extinct local, earlier budding Cinsaut-like speciality of the greater southern Rhône valley.

OJO DE LIEBRE, meaning 'hare's eye', Catalan synonym for TEMPRANILLO.

OLASZ RIZLING or **OLASZRIZLING,** once **Olasz Riesling** and still occasionally **Olaszriesling,** is the most common Hungarian name for WELSCHRIESLING, the single most planted grape in Hungary. Olasz Rizling produced around Lake Balaton is particularly prized and, in general, the warmer climate imbues Hungarian versions of this variety with quite substantial weight.

ONDARRABÍ. See HONDARRABI.

🍇 **ONDENC,** once important in Gaillac and all over South West France but now out of favour because it yields poorly and is prone to rot. During the 19th century, when it was much more popular in the greater Bordeaux region, Ondenc must have been taken to Australia, where it was called Irvine's White at Great Western in Victoria and Sercial in South Australia, only to be identified by visiting French ampelographer Paul Truel in 1976. Since then it has all but disappeared from Australian vineyards too, although rot is much less of a problem here.

OPORTO, synonym for KÉKOPORTO of which the Portuguese do not approve.

🍇 **OPTIMA,** relatively recent (1970) German crossing, of a SILVANER × RIESLING with MÜLLER-THURGAU. It ripens very early indeed, sometimes more than 10 days before Müller-Thurgau, and can notch up impressive ripeness readings, even if the wines themselves are flabby and undistinguished. It will grow on some of the poorest of sites and is therefore used, mainly in the Mosel and Rheinhessen, as a useful but ignoble booster of Prädikat level in a blend, like the more widely planted ORTEGA. Its late budding makes it popular in the Mosel-Saar-Ruwer and it is also grown in the Rheinhessen. Germany's plantings of Optima totalled 420 ha/1,000 acres in 1990 but are not expected to increase.

ORANGE MUSCAT. California rarity. See MUSCAT BLANC À PETITS GRAINS.

🍇 **ORÉMUS,** Hungarian crossing of FURMINT with BOUVIER which can in the Tokaj region produce characterful, fiery dry white varietals.

🍷 **ORION,** modern crossing in the aromatic SEYVAL BLANC mould planted to a limited extent in England.

ORMEASCO, local name for DOLCETTO on the north western coast of Italy.

ORTEGA is popular as an Oechsle-booster in German wines, especially with the blenders of the Rheinhessen. This crossing of Müller-Thurgau and Siegerrebe produces extremely full-flavoured wines that often lack acidity but can reach high must weights, if not quite as high as the equally early ripening but less widely planted OPTIMA. Varietal wines are made, and very ripe QmP Ortega is a distinct possibility even in less good vintages, but a little goes a long way. The vine does not have good disease resistance, however, and its susceptibility to coulure leaves Optima the more obvious choice for the Mosel-Saar-Ruwer. Germany's total plantings were around 1,200 ha/3,000 acres in the late 1980s and early 1990s, half of them in the Rheinhessen.

🍷 **ORTRUGO,** speciality of the hills around Piacenza in north central Italy, usually blended with MALVASIA. Anderson praises it as 'respectable'.

ÖSTERREICHER, old name for SILVANER.

OTTAVIANELLO, Apulian name for the French red grape variety CINSAUT.

P

PAARL RIESLING, old South African name for CROUCHEN.

🍷 **PADERNÃ,** Vinho Verde region name for ARINTO.

PAGADEBIT, PAGADEBITO, another name for the BOMBINO BIANCO grape enjoying a certain revival in Romagna in north central Italy, and also grown across the Adriatic in the former Yugoslavia. The name refers to the vine's reliable yields which, in theory, should allow growers to pay their debts. It is also known as Debit.

PAÏEN, Swiss name for GEWÜRZTRAMINER.

🍷 **PAIS,** the most common grape variety in Chile, identical to California's historic MISSION grape and a darker-skinned version of the CRIOLLA CHICA of Argentina. In Chile, where it is most common in the southern regions of Maule and Bío-Bío, it is also sometimes known as Negra Peruana. It may be the same as the MONICA of Spain and Sardinia.

PALOMBINA. See PIEDIROSSO.

🍷 **PALOMINO,** the sherry grape grown most nobly around Jerez in southern Spain. It is almost certainly of Andalucian origin, supposed named after one of King Alfonso X's knights. **Palomino Fino**, which once grew exclusively around Sanlúcar de Barrameda in Manzanilla country has been adopted as the most suitable variety for sherry production, as distinct from the lowlier **Palomino Basto** or **Palomino de Jerez** once widely used.

The vine is relatively susceptible to downy mildew and responds best in warm, dry soils. It has loose, generous bunches of large grapes and its yield is relatively high and regular, about 80 hl/ha (4.5 tons/acre) without irrigation. The wine produced is, typically, low in both acidity and fermentable sugars. This suits sherry producers who pick Palomino grapes at about 19° Brix and find Palomino must's tendency to oxidize no inconvenience, but for this very reason the variety tends to make rather flabby, vapid table wines, unless substantially assisted by added acid.

Of Spain's 30,000 ha/74,000 acres or so of Palomino Fino, the great majority are in sherry country around Jerez but it is also being planted in Condado de Huelva, where it is edging out the ZALEMA grape variety, and has been planted in Galicia. Outside sherry country, as in France, it is often known as Listan, or Listan de Jerez. See LISTAN for details of the declining fortunes of this variety in France. It is commonly thought to be the Perrum of the Alentejo in southern Portugal.

The country with the most Palomino planted outside Spain is South Africa, where the variety (sometimes called Fransdruif, or 'French grape'!) was the country's fourth most planted wine grape in 1994—even if its 4,700 ha/11,800 acres were a long way behind Chenin Blanc's 27,000 ha. Much South African Palomino is distilled or used for blending into basic, rather vapid table wines, but there is some attempt to increase sugar levels using canopy management techniques.

California's acreage of the variety—once wrongly identified as Golden Chasselas there—remains steady at just over 1,000 acres/400 ha, almost all of them in the Central Valley, where the wine produced is used chiefly for blending. Argentina has limited plantings of the variety and PEDRO XIMÉNEZ, another important white grape variety from southern Spain, predominates. In Australia these two varieties are not effectively distinguished and their combined total area was about 700 ha/1,750 acres in 1994, much of it grown in South Australia and used for making sherry-style fortified wines. New Zealand also grows a small amount of Palomino in its hardly ideal climate.

Cyprus has imported the Palomino vine because of its dependence on producing inexpensive copies of sherry. The variety is rated here not for the table wines it produces but for its sherries.

🍇 **PAMID,** Bulgaria's most widely planted and least interesting indigenous grape variety producing rather thin, early maturing red wines with few distinguishing marks other than a certain sweetness. It does not play a major role in bottles bound for export but, as Piros Szlanka, is planted quite extensively in Hungary, and to an even greater extent, as Rosioara, in Romania.

🍇 **PAMPANUTO, PAMPANINO,** Apulian speciality almost invariably blended with something higher in acid, although is not grown in great quantity there.

PANSA BLANCA, synonym for the Spanish white grape variety XAREL-LO, used particularly in Alella, where a pink-skinned **Pansa Rosado** is also known.

🍇 **PARDILLO,** sometimes called **Pardina,** undistinguished but very common grape in south western Spain, especially in Badajoz province on the Portuguese border. It covers more land than any Spanish grape other than the AIRÉN of La Mancha and produces generally undistinguished wines with low acidity which oxidize easily. A perfect candidate for fino-style wines.

🍇 **PARELLADA,** potentially high-quality Catalan variety widely used, with MACABEO and XAREL-LO, for the production of sparkling Cava. It is the least planted of these three varieties in the Penedès region, but became increasingly popular with vine-growers in the 1960s when Cava producers paid a premium for it. Parellada can produce a fine, refreshing wine with the crunchy crispness of green apples when grown in relatively poor soil and in cooler conditions, but has a tendency to over-produce lower-quality wine in fertile soils. It has large, loose bunches of large grapes which have good resistance to rot. The grape can also play an important part in still white wines from Cariñena, Costers del Segre and Penedès. It has been blended with CHARDONNAY and SAUVIGNON BLANC with some success, most notably in some barrel-aged examples from Torres.

🍇 **PASCAL BLANC,** almost extinct Provençal variety, very sensitive to powdery mildew and rot.

🍇 **PASCALE DI CAGLIARI,** Sardinian speciality.

PASSERINA, Marches name for BIANCAME.

🍇 **PECORINO,** Marches speciality of Italy's east coast making ever-decreasing quantities of firm, characterful wine.

🍇 **PEDRO GIMÉNEZ,** important white grape in Argentina, where it is the most planted vine other than the coarse and declining CRIOLLA CHICA and CEREZA. There were 22,600 ha/55,800 acres of Pedro Giménez in 1989, almost three-quarters of them in Mendoza province. This is the variety that underpins Argentina's white wine production and it is also found in Chile's pisco (local brandy) region. Ampelographers in Argentina believe there is no connection between this variety and the PEDRO XIMÉNEZ of Spain.

♱ **PEDRO LUIS.** See FALSE PEDRO.

♱ **PEDRO XIMÉNEZ, PEDRO JIMÉNEZ, PEDRO,** traditionally associated with Andalucia in southern Spain but now much less common than the PALOMINO Fino, in the sherry region at least. Palomino Fino is more productive and less disease-prone than Pedro Ximénez. Pedro Ximénez covers almost as much ground, however, perhaps 27,000 ha/67,000 acres, and is found all over Andalucia, Valencia, and Extremadura. It is capable of producing very ripe grapes, and is by far the dominant grape for the sherry-like wines of Montilla-Moriles. The other common fate of these thin-skinned grapes, which were traditionally dried in the sun to produce wines to sweeten sherry and other fortified blends, is to produce somewhat flabby, neutral-flavoured dry table wines, although some rich, raisiny, sweet fortified wine called Pedro Ximénez, or simply 'PX', is bottled. The variety is also grown on Spain's Canary Islands off the Atlantic coast.

In Australia, Pedro Ximénez is not distinguished from the more popular Palomino for the purposes of vineyard censuses, but the variety has been known to shine, most particularly in nobly rotten form to produce the rich, deep golden sweet wines in irrigated vineyards near Griffith in New South Wales. The variety was once confused with another Australian import from Jerez, but such Cañocazo as remains is now known as False Pedro in Australia. The vine called False Pedro by South Africa is the Andalucian variety Pedro Luis, however. California has all but dispensed with Pedro Ximénez.

♱ **PELAVERGA,** pale Piedmontese rarity making crackling strawberryish wines.

♱ **PELOURSIN,** obscure southern French vine of which DURIF was a selection. Small plantings of Peloursin, and some Durif, have been identified in north east Victoria in Australia.

PERE'E PALUMMO. See PIEDIROSSO.

PERIQUITA, Portuguese name for a parakeet given to the usefully versatile red grape variety grown all over southern Portugal, CASTELÃO FRANCÊS.

PERLAN, occasional Swiss name for CHASSELAS.

♱ **PERLE,** like WÜRZER, modern German crossing of GEWÜRZTRAMINER and MÜLLER-THURGAU. In this case, however, Gewürztraminer's rosy-hued grapes have been inherited but its extravagant perfume is more muted. It is particularly useful in Franken since its late budding protects it from spring frost damage. The wine produced is flowery but the vine's compact bunches make it an easy target for rot, which may account for its decline during the 1980s, to 200 ha/500 acres by 1990.

♛ PERRICONE, Sicilian variety planted on hardly more than 1,000 ha/2,500 acres of the island. Soft varietal wines are sometimes called by its synonym Pignatello.

♛ PERRUM, Portuguese white grape variety most commonly producing rather ordinary wines in Alentejo and thought by some authorities to be the same as the PALOMINO of Spain's sherry vineyards.

♛ PERSAN, now rare Savoie variety which can produce wines worth ageing.

♛ PETIT COURBU, characterful, traditional Gascon variety rescued from oblivion to be blended with the local ARRUFIAC, and some GROS MANSENG and PETIT MANSENG imported from Jurançon, to produce Pacherenc du Vic Bilh. Petit Courbu (Gros Courbu has been abandoned) makes rich wines which taste of lemon and honey.

♛ PETITE ARVINE, the most widely planted of the vine specialities of the Valais region of Switzerland and across the Italian border in the Valle d'Aosta producing heady, rich, perfumed wines. (See also AMIGNE and HUMAGNE BLANC.) Natural alcohol levels of 13 per cent are by no means uncommon and both dry and sweet versions are made, the latter from partially shrivelled grapes. Acid levels are good and some wines age well. Sloan has detected the scent of violets in the best examples. An unrelated Grande Arvine is being phased out and any wine labelled simply Arvine should be Petite Arvine.

♛ PETITE SIRAH, almost certainly invented, the common name in both North and South America for a variety that is less noble than and probably unrelated to the true SYRAH (of which some French growers distinguish a small-berried subvariety they call **Petite Syrah**).

Petite Sirah, which some authorities say is a name applied to a collection of several different little-known varieties often planted together, is relatively important in a wide range of warm wine regions, especially in both California and South America. In California, where there were still 2,900 acres/1,200 ha in 1992, notably in Monterey, it has been regarded as a relatively tannic, well-coloured blending partner for blowsier Zinfandels, but is increasingly valued as a varietal, basking in the somewhat refracted glory of its 'Rhône' connections. Almost all Petite Sirah vines in California are much older than the state average.

In 1990 Argentina had about the same area of Petite Sirah planted as California, more than 1,400 ha/3,500 acres in 1990, often misleadingly calling it Sirah. It is also well known in Brazil's semi-tropical climate as Petite Sirah or Petite Syrah and has produced respectable sturdy red in Mexico.

During the 1980s Petite Sirah was thought to be identical to the DURIF, sometimes spelt Duriff, a nearly extinct French vine variety, but early DNA tests suggested otherwise. Petite Sirah was first mentioned in California wine literature in the early 1880s, which also suggests that it cannot be identical to Durif which had hardly been propagated in France then. Phylloxera took its toll on early plantings of the mysterious Petite Sirah, which may at that stage have been Syrah. It has been suggested that in the early 20th century, and therefore subsequently, the name Petite Sirah was applied to a mixture, sometimes in the same vineyard, of grape varieties producing long-lived, deep-coloured red wines.

🍇 **PETIT MANSENG**, superior form of MANSENG (under which information on area planted can be found). Petit Manseng, the Jurançon vine, has particularly small, thick-skinned berries which yield very little juice (sometimes less than 15 hl/ha, although up to 40 hl/ha (2.3 tons/acre) is allowed) but can well withstand lingering on the vine until well into autumn so that the sugar is concentrated by the shrivelling process known as *passerillage* in French. It can easily reach a potential alcohol of 20 per cent without the concentration of noble rot. The variety, which is sensitive to coulure and both sorts of mildew, is used particularly for sweet, *moelleux* wines, now Gascony's Pacherenc du Vic Bilh too, and interest in its potential is growing in both California and the Languedoc.

PETIT PINEAU, synonym for ARBOIS.

PETIT RHIN, synonym for the great RIESLING grape of Germany used mainly in Switzerland.

🍇 **PETIT ROUGE**, the best dark-berried vine indigenous to Italy's Valle d'Aosta, according to Anderson.

🍇 **PETIT VERDOT**, one of Bordeaux's classic black grape varieties, no longer planted in any great quantity but enjoying a small revival in some quality-conscious vineyards. The vine ripens even later than CABERNET SAUVIGNON and is equally resistant to rot. It shares Cabernet Sauvignon's thick skins and is also capable of yielding concentrated, tannic wines rich in colour with an extra spicy note when it ripens fully, which in most Bordeaux properties happens only in riper vintages (when its extra power is least needed). Its inconveniently late ripening encouraged many producers to abandon it in the 1960s and 1970s so that total French plantings were just over 300 ha/750 acres in 1988. But, as its qualities are recognized, there has been a limited revival—not just in Bordeaux but in California too where there were 100 acres/40 ha (a third of them too young to bear fruit) of **Petite Verdot** (*sic*) in 1991, mainly in Napa. See also VERDOT.

🍇 **PHOENIX**, SEYVAL BLANC × BACCHUS crossing planted to a limited extent in England where it is treasured for its resistance to fungal diseases. Like Bacchus, it has a powerful, herbaceous aroma reminscent of elderflowers.

PICARDAN, minor, large-berried southern Rhône variety officially but usually theoretically allowed into white Châteauneuf-du-Pape. The wine is almost colourless, fairly neutral but usefully high in acidity.

PICOLIT, also written **Piccolit** and **Piccolito** in the past, speciality of Friuli in north east Italy used to make an extremely expensive sweet wine from dried grapes. Rosazzo in the Colli Orientali appears to be the variety's original home, and Picolit owes its survival to the efforts of the Perusini family of the Rocca Bernarda of Ipplis, which laboured throughout the 20th century first to identify and then to reproduce hardier clones. The Italian 1990 vineyard census found only 196 ha/486 acres.

PICPOUL or PIQUEPOUL, ancient Languedoc vine commonly encountered in Blanc, Noir, and Gris versions with the white being the most planted today, although they have frequently been mixed in the vineyard in their long history in the Midi. Piquepoul, meaning 'lip-stinger' (signifying the high acidity of its must), was cited as a useful grape at the beginning of the 17th century and was often blended with the fatter CLAIRETTE. Post-phylloxera fungal diseases, and its unremarkable yield, made it a less popular 20th century choice, except for the long-since-uprooted coastal vineyards because of the variety's good tolerance of sand.

Picpoul Noir produces alcoholic, richly scented, but almost colourless wine that is best drunk young. Although it is allowed as a minor ingredient in Châteauneuf-du-Pape and Coteaux du Languedoc, there were only 200 ha/500 acres left in all of France by the end of the 1980s.

Picpoul Blanc on the other hand, of which 500 ha remained then, has been experiencing a small revival of interest. It can provide usefully crisp blending material in the Languedoc but it is most commonly encountered as the lemony, quite full bodied Picpoul de Pinet.

PICUTENER, name for the local strain of NEBBIOLO in and near the Valle d'Aosta in the far north west of Italy.

PIEDIROSSO, Italian vine speciality of Campania, particularly on the islands of Ischia and Capri. It is also known as Palombina and Pere'e Palummo. Plantings halved during the 1980s so that there were hardly more than 1,000 ha/2,500 acres left by 1990.

PIGATO, characterful, ancient Italian white producing distinctively flavoured varietal wines in Liguria. Perhaps of Greek origin. Just a few hundred acres left.

PIGNATELLO, synonym for the Sicilian red grape PERRICONE.

PIGNEROL, old Provençal vine making a rather heavy contribution to white Bellet.

♀ PIGNOLA VALTELLINESE, speciality of the Valtellina zone in northern Lombardy.

♀ PIGNOLETTO, lively, crisp, aromatic speciality of the Bologna region in north central Italy.

♀ PIGNOLO, promising Friuli native, probably first grown in the Colli Orientali of north east Italy. The vine, whose Italian name means fussy, is a very shy bearer and for long local growers preferred more productive grape varieties until it was officially sanctioned in 1978 in the province of Udine. Production is still on a very small scale but the results suggest encouragingly high quality. The rich, full, deep-coloured wines have shown a real affinity with barrel ageing.

PINEAU, a word widely used in France as a synonym for the PINOT family of grape varieties. It seems to have been a portmanteau word for any better quality vine in medieval France (probably a reference to the pine-shape of so many bunches of grapes) but is today a word associated primarily with the Loire. The first word of a wide range of vine synonyms, sometimes various forms of PINOT but more often CHENIN, most notably as **Pineau de la Loire**.

♀ PINEAU D'AUNIS, sometimes called **Chenin Noir,** is a variety that is neither a PINOT nor a CHENIN according to Galet, but a distinct black-berried Loire vine variety associated since the Middle Ages with the Prieuré d'Aunis near Saumur. (Earlier authorities, including Viala, maintained that Chenin Blanc is a white-berried mutation of Pineau d'Aunis.) It is systematically being pulled up in favour of more fashionable or longer-living vines such as CABERNET FRANC but there were still nearly 500 ha/1,200 acres of it planted in the eastern Loire and a substantial presence in Anjou-Saumur at the end of the 1980s. The variety is one of the many sanctioned for the red and rosé appellations of Touraine and Anjou but is used only to a limited extent, mainly to bring liveliness and fruit to rosés, although in ripe years it can yield a fine red, notably in Coteaux du Loir.

PINEAU MENU, synonym for ARBOIS.

PINENC, name for FER in Madiran, South West France.

PINOT, first word of many an originally French grape name, presumably referring to the shape of the bunch in the form of a pine (*pin*) cone. Galet cites no fewer than 100 different sorts of Pinot, although most of them are synonyms. The principal true members of the Pinot family are PINOT BLANC, AUXERROIS, PINOT GRIS, MEUNIER, and PINOT NOIR, all of them related. Chardonnay is still occasionally but misleadingly called Pinot Chardonnay.

In German members of the Pinot family frequently have the word Burgunder in their German names (see SPÄTBURGUNDER, WEISSBURGUNDER, and GRAUBURGUNDER). There was a marked increase in the popularity of these grape varieties throughout the 1980s as tastes changed in favour of drier, fuller German wines.

🌱 **PINOTAGE,** South Africa's own grape, from a 1925 crossing of PINOT NOIR × CINSAUT (the latter being called, rather misleadingly, Hermitage at the time). This hardy early ripening vine was launched as a varietal as recently as the early 1960s and is currently enjoying a vogue among producers and consumers alike for its body and distinctive aroma, an exuberant, sweet, paint-like pungency (iso-amyl acetate). Pinotage is versatile and can turn out Beaujolais-style wines for drinking young and cool, or more serious oak-aged essences which are worthy of bottle ageing. The vine is productive, reliable, and reaches good levels of acidity and sugars. A little is grown in Zimbabwe and New Zealand and we can expect to see curious growers experimenting, and succeeding, with it in a wider range of countries yet.

PINOT BEUROT, ancient Burgundian synonym for the PINOT GRIS once commonly planted there.

🍇 **PINOT BIANCO,** common Italian name for PINOT BLANC, so widely grown in Italy that more wine is probably sold under this synonym than the total amount of wine labelled Pinot Blanc. In 1990 nearly 7,000 ha/17,000 acres of Pinot Bianco were counted in Italy (compared with 6,000 ha of Chardonnay and 3,400 ha of Pinot Grigio).

It is grown particularly in the north and east in Trentino-Alto Adige, Veneto, Friuli, and Lombardy although, as in Alsace but not in Germany or Austria, Pinot Grigio or PINOT GRIS enjoys higher esteem here. It was first noted in Italy in Piedmont in the early 19th century and until the mid 1980s the name Pinot Bianco was used to describe Pinot Blanc, Chardonnay, or a blend of the two. Even today there are vineyards in which both varieties grow side by side. Italians generally vinify Pinot Blanc as a high-acid, slightly fizzy, non-aromatic white for early consumption, and often coax generous yields from the vine. In Lombardy the high acid and low aroma are particularly prized by the Spumante industry.

🍇 **PINOT BLANC,** widely planted French vine, a white mutation, first observed in Burgundy at the end of the 19th century, of PINOT GRIS, which is itself a lighter-berried version of PINOT NOIR. Although its base is Burgundian, today its stronghold is in central Europe. For many years no distinction was made between Pinot Blanc and CHARDONNAY since the two varieties can look very similar to all but the keenest ampelographers. The most famous of these, Galet, identifies the distinction and cites three different strains of the variety. True Pinot Blanc is low in both vigour and productivity, while the Pinot Blanc selected for and now widely cultivated in Alsace, called Gros Pinot Blanc by Galet, is much more vigorous and productive. There is also a selection of the variety that ripens two weeks early.

PINOT BLANC

No Pinot Blanc is notable for its longevity, nor its piercing aroma; its scent arrives in a cloud. Most wines based on Pinot Blanc are also relatively full bodied, which has undoubtedly helped reinforce the confusion with Chardonnay, not only in Burgundy but also in north-east Italy. Although Chardonnay dominates white burgundy, Pinot Blanc is technically allowed into several white (and some red) Burgundian appellations, but is no longer grown in any quantity in Burgundy.

Alsace is Pinot Blanc's current French stronghold, although even here it is less important in terms of total area planted than Riesling, Gewürztraminer, Silvaner, or even the related white AUXERROIS with which it is customarily blended in Alsace, to be sold as 'Pinot Blanc'. In Luxembourg on the other hand, the higher acidity of Pinot Blanc makes it much less highly regarded than Auxerrois.

While in Alsace it is regarded as something of a workhorse (and sometimes called Clevner or Klevner), it has been generally held in higher esteem by the Germans, who have a rather greater area planted (although much less in total than they have of the Pinot Gris they call Ruländer or Grauburgunder) and call it Weissburgunder or Weisser Burgunder. They have valued it for its apparent similarity to the world-famous Chardonnay and its ability to accumulate sugar in ripe grapes, even at relatively high yields. Planted mainly but not exclusively in Eastern Germany, Pfalz, and Baden, it is a popular vehicle for fuller, drier wine styles designed to be drunk with food and has been keenly adopted as a suitable vehicle for barrel ageing.

As PINOT BIANCO it is a popular dry white in Italy but it is in Austria that, as Weissburgunder, the variety reaches its greatest heights, and certainly its greatest ripeness. Accounting for almost 4 per cent of the country's total vineyards, it is grown in all regions, although Riesling is more prized in the Wachau. As a dry white varietal, Weissburgunder is associated with an almond-like scent, relatively high alcohol, and an ability to age, but it has achieved its greatest Austrian glory in ultra-rich Beerenauslese or even Trockenbeerenauslese form, thanks to noble rot effects in the vineyards of Burgenland.

Pinot Blanc is widely disseminated over Eastern Europe. In Slovenia, Croatia, and Vojvodina it is widely grown and may be called Beli (White) Pinot.

Vine-growers in the New World recognize that Pinot Blanc has lacked Chardonnay's glamour, but in Monterey there were still several hundred acres of it, some of the highest-priced being Chalone's Chardonnay-like examples. Older vines bearing this name are almost certainly not Pinot Blanc but the Muscadet grape MELON (although newer plantings, typically in the cool Carneros for sparkling wine production, are true Pinot Blanc). The fact that for long within California no great distinction was noticed between the wines made from true Pinot Blanc and Melon adds further weight to the thesis that Melon was originally a Burgundian variety.

Elsewhere in the New World, Pinot Blanc is largely ignored in favour of the most famous white wine grape.

PINOT BLANCO, common misnomer for CHENIN BLANC in Mexico and South America.

PINOT CHARDONNAY, erroneous and misleading synonym for CHARDONNAY, adopted at a time when Chardonnay was believed to be a white mutation of PINOT NOIR.

PINOT DE LA LOIRE, misleading occasional synonym for CHENIN BLANC, also known as Pineau de la Loire.

🍇 **PINOT GRIGIO**, common Italian name for PINOT GRIS and, as such, probably the name by which the grape is best known to many wine drinkers. There were about 3,500 ha/8,600 acres of Pinot Grigio vineyard in Italy in 1990 (much less than the area planted with PINOT BIANCO, for example). Most of these plantings were in the north east and specifically in Friuli, where it produces some of the most admired wines of Collio as well as a sea of reasonably undistinguished dry white with low aroma and probably the most noticeable acidity of any of the world's Pinot Gris. The Italian tendency is to pick the grapes before the variety's characteristically rapid loss of acid at full ripening. The variety is also grown widely in Lombardy, where it is gobbled up by the sparkling wine industry. The variety can also be found as far south as Emilia-Romagna and is planted in Alto Adige—although here, as in all Germanic areas, Pinot Bianco is favoured.

🍇 **PINOT GRIS**, widely disseminated vine that can produce soft, gently perfumed wines with more substance and colour than most whites, which is what one might expect of a variety that is one of the best-known mutations of PINOT NOIR. If Pinot Noir berries are purplish-blue and the berries of the related PINOT BLANC are greenish-yellow, Pinot Gris grapes are anything between greyish blue and brownish pink—sometimes on the same bunch. In the vineyard this vine can easily be taken for Pinot Noir for the leaves are identical and, especially late in a ripe year, the berries can look remarkably similar. At one time Pinot Gris habitually grew in among the Pinot Noir of many Burgundian vineyards, adding softness and sometimes acidity to its red wine. Even today, as Pinot Beurot, it is sanctioned as an ingredient in most of Burgundy's red wine appellations and the occasional vine can still be found in some of the region's famous red wine vineyards. It was traditionally prized for its ability to soften Pinot Noir musts but older clones have a tendency to yield very irregularly.

There also remain small pockets of the variety in the Loire where it is often known as Malvoisie (although even in such a small appellation as Coteaux d'Ancenis both Malvoisie and Pinot Beurot are officially allowed as part of the appellation name). It can produce perfumed, substantial wines in a wide range of different sweetness levels. It is also known as Malvoisie in the

Valais in Switzerland, where it can also produce full, perfumed, rich whites.

But within France, Alsace is where Pinot Gris (here traditionally but mysteriously known as Tokay) is most revered, and with good reason. It may be less commonly planted than the other members of Alsace's noble triumvirate Riesling and Gewürztraminer, but it fulfils a unique function as provider of super-rich, fairly dry, wines that can be partnered with even the heartiest savoury foods without the distraction of too much aroma.

As with Pinot Blanc, however, much more Pinot Gris is planted in both Germany and Italy than France. (See PINOT GRIGIO for details of Italian Pinot Gris.) In Germany it is usually known as RULÄNDER or, if the wine is dry, Grauburgunder (or occasionally Grauer Riesling, Grauer Burgunder, or even Grauklevner).

Like Pinot Blanc, it is widely planted not just in Austria but in Slovenia, Moravia, and particularly Romania where, on 1,600 ha/4,000 acres of vineyard, it is known both as Pinot Gris and Ruländer, and Hungary where it is revered as SZÜRKEBARÁT. Pinot Gris is also grown in both Russia and Moldova.

Pinot Gris's impact on the New World is distinctly limited although improved clonal selection has precipitated a renewal of enthusiasm in California where its 'scented Chardonnay' style may have potential, and on the South Island of New Zealand, albeit on a small scale.

The variety is also much admired for its weight and relatively low acidity in Luxembourg.

PINOT LIÉBAULT, unusual and slightly more productive Burgundian selection of PINOT NOIR first identified in Gevrey by A. Liebault in 1810, according to Galet.

PINOT MEUNIER. See MEUNIER.

PINOT NERO, Italian for PINOT NOIR. The variety is quite widely planted in the north east of the country and in Lombardy, and demand for Spumante doubled its area to a total of about 3,500 ha/8,600 acres in the 1980s, but few examples show great intensity of flavour.

 PINOT NOIR, the great red burgundy grape which gives its name to the NOIRIEN family of grape varieties. Unlike CABERNET SAUVIGNON, which can be grown in all but the coolest conditions and can be economically viable as an inexpensive but recognizably Cabernet wine, Pinot Noir demands much of both vine-grower and wine-maker. It is a tribute to the unparalleled level of physical excitement generated by tasting one of Burgundy's better reds that such a high proportion of the world's most ambitious wine producers want to try their hand with this capricious vine. Although there is little consistency in its performance, Pinot Noir has been transplanted to almost every one of the world's wine regions, except the very hottest where it can so easily turn from essence to jam.

If Cabernet produces wines to appeal to the head, Pinot's charms are decidedly more sensual and more transparent. The Burgundians themselves refute the allegation that they produce Pinot Noir; they merely use Pinot Noir as the vehicle for communicating local geography, the characteristics of the individual site on which it was planted. Perhaps the only characteristics that the Pinot Noirs of the world could be said to share would be a certain sweet fruitiness and, in general, lower levels of tannins and pigments than the other 'great' red varieties Cabernet Sauvignon and SYRAH. The wines are decidedly more charming in youth and evolve more rapidly, although the decline of the very best is slow. Pinot can taste of raspberries, strawberries, cherries, and violets in youth, while ageing through more autumnal or spicy scents to something considerably more gamey after many years in bottle.

Part of the reason for the wide variation in Pinot Noir's performance lies in its genetic make-up. It is a particularly old vine variety, in all probability a selection from wild vines made by man at least two millennia ago. There is some evidence that Pinot existed in Burgundy in the fourth century AD. Although Morillon Noir was the common name for early Pinot, a vine called Pinot was already described in records of Burgundy in the 14th century and its fortunes were inextricably linked with those of the powerful medieval monasteries of eastern France and Germany.

Pinot Noir has for long been grown in Burgundy, but it is particularly prone both to mutate (as witness PINOTS BLANC, GRIS and MEUNIER) and degenerate, as witness the multiplicity of Pinot Noir clones available even within France.

Galet notes that no fewer than 46 Pinot Noir clones (as opposed to 34 of the much more widely planted Cabernet Sauvignon) are officially recognized within France—and that in the 1980s only Merlot cuttings were more sought after from French nurseries than those of Pinot Noir. It is thus possible to choose a clone of Pinot Noir specially chosen for its productivity, for its resistance to rot and/or for its likely ripeness (which can vary considerably). Most selection work has been done in Burgundy and Champagne and it has been the Champenois who have selected particularly productive clones. It is generally agreed that a major factor in the lighter colour and extract of so much red burgundy in the 1970s and 1980s was injudicious clonal selection, resulting perhaps in higher yields but much less diversity and concentration in the final wine. The most planted clone in Burgundy is 115 but 114 is more highly regarded and the most reputable producers of all tend to have made selections from their own vine population. The clone called Pommard is well distributed in the New World, as has been one named after the Wädenswil viticultural station in Switzerland. In general the most productive clones, which have large-berried bunches, are described as Pinot Droit for the vines' upright growth, while Pinot Fin, Pinot Tordu, or Pinot Classique grows much less regularly but has smaller berries with thicker skin.

Inasmuch as generalizations about a vine variety with so many different forms are possible, Pinot Noir tends to bud early, making it susceptible to spring frosts and coulure. Damp, cool soils on low-lying land are therefore best avoided. Yields are theoretically low, although too many Burgundians disproved this with productive clones in the 1970s and early 1980s. The vine is also more prone than most to both sorts of mildew, rot (grape skins tend to be thinner than most), and to viruses, particularly fanleaf and leafroll. Indeed it was the prevalence of disease in Burgundian vineyards that precipitated the widespread adoption of clonal selection there in the 1970s.

Pinot Noir seems to produce the best quality wine on limestone soils and in relatively cool climates where this early ripening vine will not rush towards maturity, losing aroma and acidity. Pinot Noir can be difficult to vinify, needing constant monitoring and fine tuning of technique according to the demands of each particular vintage. The trick is to leech colour and flavour but not too much tannin from these relatively thin-skinned grapes, although some wine-makers keep some stems in the fermentation vat in a soft year when they fear tannin levels are dangerously low.

Pinot Noir is planted throughout eastern France and has been steadily gaining ground from less noble varieties so that by 1988 its total area of French vineyard was 22,000 ha/54,000 acres, twice as big as the total area planted with Pinot Meunier but less than Syrah's total—and considerably less than the total planted with Burgundy's other red vine variety GAMAY because of the vast extent of Beaujolais in comparison with the famous Côte d'Or Burgundian heartland.

The Côte d'Or was once the wine region with the biggest single area of Pinot Noir, but the extension of the Champagne region in the 1980s meant that by the end of the 1980s, more Pinot Noir went into champagne than into red burgundy. In 1988 there were 6,000 ha/15,000 acres of Pinot Noir in the Côte d'Or. Even in the Greater Burgundy region, Pinot Noir is rarely blended with any other variety, except occasionally with Gamay in a Bourgogne Passetoutgrains and, increasingly, to add class to a Mâcon. At one time, Pinot Gris would be planted randomly in the same vineyard as Pinot Noir and the varieties would be vinified together. The red wines of Burgundy can vary from deeply coloured, tannic, oak-aged mouthfuls that demand long bottle age to acidic dark rosés that should be drunk as young as possible. The best grands crus are intense, fleshy, vibrant, fruity wines with structure but oak influence that is never obvious.

Pinot Noir is also gaining ground in the Côte Chalonnaise and, to a lesser extent, the Mâconnais, typically at the expense of Gamay, which is in decline in both of Burgundy's subregions between the Côte d'Or and Beaujolais. Pinot Noir occupied a full quarter of all Mâconnais vineyard in 1988 and the Chalonnaise reds of Mercurey, Givry, and Rully have shown that they can often deliver a more consistent level of wine-making than more expensive appellations to the north, even if the fruit quality may be slightly more rustic.

Pinot Noir is the favoured black grape variety in northern Burgundy too for light, fruity reds and rosés of Irancy, Côtes de Toul, and the wines of Moselle—often called *vins gris*.

Pinot Noir is also planted, to a limited but increasing extent, in the most easterly vineyards of the Loire and its tributaries, most notably to make red and pink Sancerre but also in Menetou-Salon, and St-Pourçain, and is technically allowed in an array of the Loire's VDQS wines. It was taken to the vineyards of the Jura and Savoie from Burgundy centuries ago but it is usually blended with local grapes and is rarely seen as a single varietal. It is rarely encountered in the south or west of France, the domains of Syrah and Cabernet Sauvignon respectively, but there are limited plantings in Minervois and Limoux with some intriguing, if atypical, results.

In Alsace, where it has been an important vine since the early 16th century, Pinot Noir is effectively the only black grape capable of producing quite deep-coloured, perfumed, slightly sweet reds in the ripest vintages. In cooler years it produces deep-pink wines, often with a similar smoky perfume to a white Pinot Gris, that can be reminiscent of German Spätburgunders, even down to the whiff of rot in the wettest vintages.

Germany's rediscovered interest in connoisseurship of the 1980s resulted in a marked increase in demand for the nation's noblest red so that by 1990 (Blauer) Spätburgunder was Germany's fourth most planted vine after RIESLING, MÜLLER-THURGAU, and SILVANER with more than 5,500 ha / 13,500 acres of vineyard, nearly 60 per cent more than a decade previously. An increasing proportion of German Pinot Noir is made in the image of the best Burgundian prototype to be completely dry, deep coloured, and well structured, thanks to much lower yields, longer maceration, and sometimes maturation in small oak barrels. There are ambitious producers in the Rheingau, Rheinpfalz, and Baden, where two-thirds of Germany's Pinot Noir is grown. Others, however, continue the tradition of making pale pink wines, often with notable sweetness such as Weissherbst. Such wines have been the traditional specialities of Assmannshausen in the Rheingau and the Ahr. Seriously sweet Spätburgunder is occasionally made as a late picked Beerenauslese in the Rheingau and can command the dizzy prices associated with rarities.

It is fair to say, however, that whereas Cabernet Sauvignon is often associated with the flavours of oak, Pinot Noir is often encountered with more than its fair share of sweetness, especially in inland Europe, as though over-chaptalization had been adopted as an alternative to true ripeness. Austria's Blauer Spätburgunder, for example, can taste sweet and oddly viscous unless from one of its most skilled practitioners. Total Austrian plantings are limited, however, and the native grape ST LAURENT, sometimes confusingly called Pinot St Laurent for its Pinot Noir-like soft fruitiness, is much more common, as it is in Slovakia.

Pinot Noir is spread widely, if not thickly, throughout the vineyards of eastern Europe although its name is usually some variant on the local word for Burgundian. There are plantings of Burgundac Crni in parts of Croatia, in Serbia where it can be quite successful, and in much paler form, in Kosovo. It is also grown to a limited extent in Bulgaria, Georgia, Azerbaijan, Kazakhstan, and Kyrghyzstan. Romania has about 1,200 ha / 3,000 acres and its own subvariety known as Burgund Mare which can be cheap and syrupy. Moldova grows the substantial total of 7,000 ha / 17,600 acres of Pinot according to official statistics.

Elsewhere in Europe the finicky nature of the Pinot Noir vine has set a natural limit on its spread. Miguel Torres has managed to coax some flavours reminiscent of red burgundy from his small Pinot Noir vineyard in Catalonia. But the vine is relatively important in Switzerland, notably as Klevner around Zürich and blended with Gamay in the ubiquitous Dôle, and in some cooler Italian wine regions where it is known as PINOT NERO.

It was wine producers in the New World, however, who turned the full heat of their ambitious attentions on Pinot Noir in the late 1980s and early 1990s. Some even relocated their wineries many hundreds of miles in order to be closer to sources of suitably cool climate Pinot Noir fruit. Although for many years Oregon, with its often miserably cool, wet climate was popularly supposed to provide America's only answer to red burgundy, a flock of seriously fine Pinot Noirs emerged in the late 1980s from cooler parts of California such as Russian River, Carneros, Chalone, and the Gavilan mountains of San Benito and Santa Maria. By 1992 there were nearly 10,000 acres / 4,000 ha of Pinot Noir in California, notably in fog-cooled Carneros, where the variety is also valued as an ingredient in champagne-like sparkling wines.

The variety called PINOT ST GEORGE in California is unrelated to any known Pinot, while that called Gamay Beaujolais (of which more than 1,000 acres remained in 1992) is a clone of Pinot Noir, although not one embraced by the most ambitious producers of California Pinot Noir.

Outside California and Oregon (where Pinot Noir is sometimes picked a good six weeks after it is in California), the variety has no established American outpost of great reputation. Washington State's plantings of 300 acres were in decline in the early 1990s, making way for Merlot. There are pockets of Pinot Noir in Canada's cool vineyards, however, and they are producing increasingly successful wines.

As **Pinot Negro,** it is known in most Argentine provinces where vines are grown but the climate is too hot and irrigation too commonplace to produce wines of real quality, as in most of the rest of South America, although Chile had 200 ha planted in 1994 and is producing some increasingly successful results.

Across the Atlantic in South Africa progress has been slowed by the quality of clones commonly planted but the potential is certainly there in the coolest, southernmost vineyards, even if the quantity of true Pinot Noir planted is still minute compared with total area of South Africa's signature black grape variety PINOTAGE, of which Pinot Noir is an unlikely parent.

Plantings of Pinot Noir in Australia and New Zealand on the other hand rose substantially in the early 1990s as an increasing number of producers mastered the art of replicating the Pinot Noir made in both Burgundy and Champagne. Martinborough, Canterbury, and Central Otago have produced particularly impressive results in New Zealand. Total Australian plantings of Pinot Noir were 1,100 ha/2,700 acres in 1991, of which almost a third were still too young to bear fruit—a signal in part of the success of the Australian sparkling wine industry but also of the determination of Australian wine-makers to overcome the disadvantages of their warm climate and join the world's fine wine-making club. Areas with proven success as producers of good quality Pinot Noir as a still red varietal are Geelong, Yarra, and Mornington Peninsula, all relatively cool areas around Melbourne in Victoria.

It should not surprise those familiar with the English climate that some growers have found success with Pinot Noir in its southern counties, although this really is a marginal exercise.

🍷 PINOT ST GEORGE, California variety in sharp decline which Galet believes to be NÉGRETTE.

PIROS SZLANKA. See PAMID.

🍇 PLANTA FINA (DE PEDRALBA), grown in Valencia, south eastern Spain to make sturdy, aromatic wines.

🍇 PLANTA NOVA, another undistinguished Spaniard planted on about 1,800 ha/4,500 acres of Valencia and Utiel-Requena.

🍷 PLANTET, the Loire's most popular French hybrid. France's total area shrank from more than 26,000 ha/64,000 acres to less than 1,000 ha between 1968 and 1988, almost all of it in the Loire, although at one point it was grown all over France's northern wine regions. Its chief attributes are its productivity and its ability to crop regardless of the severity of the winter and spring frosts (although New York state winters have proved too harsh for it). It has at least one SEIBEL parent.

🍷 PLAVAC MALI, responsible for dense wines all along the Dalmatian coast and on many of the Adriatic islands in Croatia. Mali means 'small' and a white grape variety called simply **Plavac** is also known, and results in equally heady wines. Both varieties thrive on sandy soil. Plavac Mali produces wines high in tannins, alcohol, and colour which can, unusually for red wines from what was Yugoslavia, age well. Postup and Dingač are two of the better-known reds made from Plavac Mali. Some observers insist that it is identical to ZINFANDEL and PRIMITIVO.

PLAVAI

🍇 **PLAVAI**, late ripening vine native to Moldova and widely
planted throughout eastern Europe and the ex-Soviet Union. In
Moldova it is also known as Belan and Plakun; in Romania it is
called Plavana; in Austria Plavez Gelber; in Hungary Melvais, in
the Krasnodarski region of Russia Belan or Oliver; in Ukraine Bila
Muka or Ardanski; and in Central Asia Bely Krugly. It is used in
Russia for both table wines and brandy.

🍇 **PLAVINA**, northern Croatian grape which may also be called
Brajdica. Sometimes used as a synonym for PLAVAC MALI.

🍇 **POLLERA NERA**, ancient Ligurian and north west Tuscan
vine.

🍇 **PONTAC**, historic South African TEINTURIER of mysterious
origins producing very small quantities of rough wine today.

🍇 **PORTAN**, like CALADOC and CHASAN, crossing made by
ampelographer Paul Truel, a Galet disciple, in the Languedoc. In
this case he crossed GRENACHE Noir and Portugais Bleu (Blauer
PORTUGIESER) to develop a Grenache-like variety that would ripen
even in the Midi's cooler zones. It is grown to a strictly limited
extent but is allowed into Vin de Pays d'Oc.

🍇 **PORTUGIESER** or **BLAUER PORTUGIESER**, common in
both senses of that word in both Austria and Germany, its name
suggesting completely unsubstantiated Portuguese origins. The
vigorous, precocious vine is extremely prolific, easily producing
160 hl/ha/(9 tons/acre), thanks to its good resistance to coulure,
of pale, low-acid red, often given a sweet character by enthusiastic
chaptalization.

Blauer Portugieser is synonymous with dull, thin red in Lower
Austria, where it is particularly popular with growers in Pulkautal,
Retz, and the Thermenregion. It covers five times more vineyard
area than Blauburgunder (PINOT NOIR) and is the country's second
most planted dark-berried vine variety after Blauer ZWEIGELT. Such
wines are rarely exported, with good reason, and are rarely
worthy of detailed study.

Germany's everyday black grape variety, Portugieser was
overtaken in terms of total area planted by Spätburgunder (Pinot
Noir) in the 1970s, although in the early 1990s it was enjoying a
renaissance of popularity with growers determined to satisfy the
German thirst for red wine regardless of its quality. In 1990 there
were more than 4,000 ha/10,000 acres of Portugieser in Germany,
half of them in the Pfalz, where a high proportion of Portugieser
is encouraged to produce vast quantities of pink Weissherbst. It
also plays an important role in the Ahr region, famous for its reds,
many of which depend on Portugieser as much as the
Spätburgunder for which the region is more famous. (In 1990, the
Ahr's plantings of Portugieser were nearly half those of the nobler
variety.)

The variety is so easy to grow, however, that it has spread

throughout central Europe. It is probably the KÉKOPORTO in Hungary and Romania and is grown in northern Croatia as **Portugizac Crni**, or **Portugaljka**.

At one time it grew all over South West France, where it is known as **Portugais Bleu**. Its once powerful influence in Gaillac is only slowly declining.

🍇🍇 POULSARD, sometimes called PLOUSARD, Jura rarity, one of the region's dark grapes adapted over the centuries to its very particular climate and soils. Its large, long, thin-skinned grapes give only lightly coloured wine that is distinguished by its perfume. It is so low in pigments it can make white wines, and particularly fine rosés which may be left on the skins for as long as a week without tainting the wine too deeply. It can also add aroma to the region's TROUSSEAU and, increasingly, PINOT NOIR. It is also grown to a very limited extent in nearby eastern French wine region Bugey.

PRENSAL. See MOLL.

PRESSAC, Bordeaux right bank name for the red grape variety Cot or MALBEC.

🍇 PRIETO PICUDO, unusually musky grape grown in a large area totalling about 6,700 ha/16,700 acres, mainly around the city of León in north central Spain. Wines are light in colour but very distinctive, even if they are not officially embraced by the Denominación de Origen system.

PRIMATICCIO, occasional name for Italy's MONTEPULCIANO.

🍇 PRIMITIVO, southern Italian grape, grown so extensively, especially in Apulia, that its 17,000 ha/42,000 acres made it Italy's ninth most planted vine variety in 1990 (just behind Apulia's even more popular NEGROAMARO). It makes deeply coloured, well-structured, spicy wines quite high in alcohol, though not as headily so as Negroamaro. The international spotlight fell on it in the late 1970s, when a similarity was noted with California's popular ZINFANDEL. Subsequent DNA tests confirm this theory but there has so far been no definitive proof that the variety originated in Apulia. Wine-making techniques (and equipment) are considerably less sophisticated in southern Italy than in northern California, but the variety can produce excitingly characterful wines.

PRINÇ, synonym for TRAMINER occasionally used in Moravia.

PROCANICO, Umbrian and possibly superior subvariety of TREBBIANO.

🍇 PROKUPAC, robust vine grown all over Serbia, where the strong wine it produces, in good quantity, is often blended with more international vine varieties. It is also grown in Kosovo and Macedonia. Within its native land it is often made into a dark rosé. Its stronghold is just south of Belgrade, where some argue it is related to SYRAH.

🍇 **PROSECCO,** late ripening vine native to the Friuli region in north east Italy, responsible for a popular, lightly fizzing wine of the same name, sometimes called Prosecco di Conegliano Valdobbiane. It typically has a slightly astringent or bitter finish. A total of about 7,000 ha/17,000 acres of Prosecco were planted in Italy in the early 1990s.

The variety is also known, to a very limited extent, in Argentina.

🍇 **PROVECHON,** Spanish grape grown in Aragon.

🍇 **PRUGNOLO GENTILE,** local name for a clone of SANGIOVESE distinguished in Montepulciano, of Vino Nobile fame, in early 18th century Tuscany.

PX, common abbreviation for the Spanish PEDRO XIMÉNEZ.

R

🍇 **RABO DE OVELHA,** grape grown all over Portugal taking its name from the 'ewe's tail' shape of its bunches. Noted more for alcohol than subtlety.

🍇 **RABOSO,** grown in the Veneto region of north east Italy, notably on the flat Piave valley floor. Its name may be inspired by the angry, or *rabbioso*, reaction to its very acid, astringent impact on the palate, whether as grape or wine. Its thick skins give it excellent resistance to disease and rot, but Raboso unfortunately rarely boasts the alcohol needed to compensate for its astringency and can therefore taste extremely austere in youth. Stalwart defenders insist that gentle, sympathetic wine-making is the key, and that Raboso could be the Veneto's answer to the NEBBIOLO of Piedmont or the SANGIOVESE of Tuscany. The reputation and price level of Raboso make it difficult to justify investment in this, however, and the rating reflects current reality. The Italian vineyard survey of 1990 found that total plantings of the **Raboso Piave** vine had almost halved since 1982, to 2,000 ha/5,000 acres.

Raboso is also planted, to an extremely limited extent, in Argentina, presumably taken there by Italian immigrants.

RAFFIAT, alternative name for ARRUFIAC.

🍇 **RAISIN BLANC,** obscure South African speciality which represents only a tiny proportion of all vines planted and makes undistinguished wines.

RAJNAI RIZLING, Hungarian name for RIESLING.

🍇 **RAMISCO,** the vine of Portugal's shrinking Colares region, and therefore, thanks to Colares's sandy, phylloxera-resistant soils, probably the only *vinifera* variety never to have been grafted. The few hundred acres left produce wines of real character that are extremely tannic in youth.

RANINA, Slovenian synonym for BOUVIER.

🍇 **RÄUSCHLING,** most commonly planted today in German-speaking Switzerland, where it can produce fine, crisp wines. In the Middle Ages it was widely cultivated in Germany, particularly the Baden region. The KNIPPERLÉ of Alsace is an early ripening subvariety.

REBULA, Slovenian name for RIBOLLA.

🍇 **REFOSCO,** vine making usefully vigorous wine in the Friuli region of north east Italy. The finest subvariety is known in Friuli as **Refosco dal Peduncolo Rosso** because of its red stem. It may have been grown here since classical times. The resulting wine is deeply coloured and lively with plummy flavours and a hint of almonds, a medium to full body, and noticeable acidity, which can be difficult to control or moderate, the variety being a notoriously late ripener. Refosco has the advantage of good resistance to autumn rains and rot.

There was a significant return of interest in Friuli's Refosco in the 1980s, and much greater care was taken in its cultivation and vinification in an effort to improve the wine's quality, including variable experiments with small barrels. The most promising zone for Refosco is Colli Orientali. Others include Grave del Friuli and Lison-Pramaggiore, and outside Friuli, Latisana, Aquileia, and Carso, where the local subvariety is known as Terrano, a reference to the variety's name Teran just across the border in Slovenia and Croatia, where it produces very similar wines.

Burton Anderson cites Cagnina as its synonym in Romagna, where it may have been planted by the Byzantines. Some ampelographers believe it is identical to the MONDEUSE NOIRE of France's Savoie, although this is denied by some French nurserymen. As Refosco, the variety is grown in California to a very limited extent.

🍇 **REGNER,** increasingly popular German crossing made in 1929 from a white table grape Siedentraube and GAMAY. Nearly 170 ha/420 acres were planted by 1990, mainly in the Rheinhessen. It buds and ripens very early and can reach impressive must weights, even if at the expense of acidity. This is not in principle a wine to drink as a varietal—although tests in England have been successful.

🍇 **REICHENSTEINER,** modern German crossing whose creator Helmut Becker maintained was the first European Community crossing with French, Italian, and German antecedents. In 1978 he developed this crossing of MÜLLER-THURGAU with a crossing of the

RÈZE

French table grape MADELEINE ANGEVINE and the Italian Early
Calabrese. Its antecedents are hardly noble and both wine and
vine most closely resemble its undistinguished German parent,
but Reichensteiner, with its looser bunches, is less prone to rot
and well-pruned plants can ripen well in good years. There were
more than 300 ha/750 acres planted in Germany in 1990 with
more than half, as is so often the case with the newer crossings, in
Rheinhessen. The variety, named after a Rhineland castle like
EHRENFELSER, is the third most planted variety in England, where
its wines have enjoyed some success. In New Zealand too it has
its followers among growers in damper areas.

🍇 **RÈZE**, extremely rare Swiss speciality responsible for the
sherry-like *vin des glaciers*. Almost extinct but a little is planted
near Sierre.

RHEIN RIESLING, or **RHEINRIESLING**, common synonym
in German-speaking countries for the great white RIESLING grape
variety of Germany.

RHINE RIESLING, common synonym for the great white
RIESLING grape variety of Germany, colloquially abbreviated to
Rhine in Australia, where it is most common.

🍇 **RIBOLLA**, variety also known as **Ribolla Gialla** to distinguish
it from the less interesting **Ribolla Verde**, best known in Friuli in
north east Italy, but several thousand ha are also grown, as Rebula,
in Slovenia. It is almost certainly the ROBOLA of the Greek island
of Cephalonia. It dates from at least the 13th century in Friuli, but
lost ground when French varieties were planted with enthusiasm
in the early 20th century. In the mid 1990s Ribolla accounts for less
than 1 per cent of all the white DOC wines of Friuli. Rosazzo and
Oslavia are generally considered the two classic areas for Ribolla
Gialla. The wine is light, delicate, floral, and sometimes lemon-
scented. Some attempts at new oak ageing have been made in
recent years, particularly in Oslavia.

Ribolla Nera is the SCHIOPPETTINO grape.

🍇 **RIESLANER**, SILVANER × RIESLING crossing grown to a very
limited extent in Germany's Franken region, where, provided it
can reach full ripeness, it can produce wines with race and currant
fruit. Fewer than 40 ha/100 acres were planted in 1990 and its late
ripening is likely to reduce this total still further.

RIESLER, confusing Austrian synonym for WELSCHRIESLING.

🍇 **RIESLING**, a great white wine grape associated most obviously
but by no means exclusively with Germany. Riesling can claim to
be the finest white grape variety in the world on the basis of the
longevity of its wines and their ability to transmit the
characteristics of a vineyard without losing Riesling's own
inimitable style. Its name has been considerably debased by being

applied to a wide range of white grape varieties of varied and often doubtful quality, the ultimate backhanded compliment.

In the late 19th and early 20th centuries, German Riesling wines were prized, and priced, as highly as the great red wines of France. Connoisseurs knew that, thanks to their magical combination of acidity and extract, these wines could develop for decades in bottle, regardless of (often low) alcoholic strength. Riesling is made at all levels of sweetness, and it is indubitable that the high proportion of late 20th century German wines that have been far too low in extract and, for many consumers, too high in residual sugar, has damaged Riesling's reputation. The average residual sugar of Riesling made everywhere is gradually declining, but the variety will surely always be distinguished for its ability to produce great sweet wines, whether they be the cold weather speciality Eiswein or ice wine, or the late harvest, often nobly rotten, Beerenauslese and Trockenbeerenauslese and their counterparts outside Germany. Riesling's high natural level of natural fruity grape acid provides it with a much more dependable counterbalance to high residual sugar than, for example, the SÉMILLON grape of Sauternes.

Riesling wine, wherever produced, is also notable for its powerful, rapier-like aroma, variously described as flowery, steely, honeyed, and whichever blend of mineral elements is conveyed by the individual vineyard site. This distinctive aroma (which can take on petrol overtones with age), usually experienced in conjunction with Riesling's natural race and tartness, is particularly high in monoterpenes (particularly aromatic flavour compounds), 10 to 50 times higher, for instance, than WELSCHRIESLING, a less noble and unrelated white grape variety prevalent in central Europe which, much to German fury, has borrowed the word Riesling for many of its aliases ('Riesling Italico', for example).

Viticulturally, true Riesling (often called White, Rhine, or Johannisberg Riesling) is distinguished by the hardness of its wood, which helps make it a particularly cold-hardy vine, making it a possible choice for relatively cool wine regions, even if it may need the most favoured, sheltered site in order to ripen fully and yield economically. So resistant is it to frost that winter pruning can begin earlier than with other varieties. Its growth is vigorous and upright, and this is a top quality variety which seems able to produce yields of 60 or 70 hl/ha (4 tons/acre), without any necessary diminution of quality. Its compact bunches of small grapes make it relatively prone to rot, and coulure can be a problem, but its chief distinction in the vineyard is its late budding and, in German terms, late ripening. (Compared with other international varieties it is relatively early ripening, but ripens very much later than MÜLLER-THURGAU and the host of other German crossings commonly planted in modern Germany.)

In the northern hemisphere it is often not picked until mid October or early November (and even later when sweet wines are made). Riesling whose ripening is accelerated by being grown in warmer regions can often taste dull; a long, slow ripening period suits Riesling and manages to extract maximum flavour, while maintaining acidity. Thus, many of Germany's (and therefore most of the world's) most admired Rieslings are grown on particularly favoured sites in cooler regions such as the Mosel-Saar-Ruwer.

ELBLING, SILVANER, and RÄUSCHLING seem to have a longer history in Germany but Riesling has certainly been grown there for at least 400 years. Today, partly thanks to the efforts of a special centre for the clonal selection of Riesling at Trier, the German vine-grower can choose from more than 60 clones of Riesling, of which one of the more controversially perfumed is the N90 used by such innovative growers in the Pfalz region as Müller Catoir and Lingenfelder. (French-certified clones of Riesling numbered precisely one, 49, in 1990, on the other hand.)

It takes effort and a particularly good site to ripen Riesling in Germany, however. Throughout the 20th century Riesling has been systematically displaced by Müller-Thurgau and many even newer, flashier crossings. By 1980, Riesling's 18,900 ha represented barely 19 per cent of all German vineyards. Since then, however, there has been some recognition of Riesling's superiority, helped by some harsh winters which severely tested Müller-Thurgau. By 1990, there were well over 21,000 ha / 52,000 acres of Riesling in about-to-be-unified West Germany, which represented about 21 per cent of the total.

The exception to this recent revival, and indeed a brake on it, is the Riesling showcase of the northerly Mosel-Saar-Ruwer which is the home of about a third of all Germany's Riesling. The variety represented 80 per cent of Mosel-Saar-Ruwer plantings in 1964 and only 54 per cent in 1990, because of the philosophy of quantity over quality which has prevailed in the large bottling operations centred in this region. The finest estates here are without exception dedicated to Riesling and plant the variety on their finest sites to the exclusion of all else. Some would argue that Riesling finds its finest expression on the steep banks of the Mosel and its Saar and Ruwer tributaries, ideally with a 30 per cent gradient to attract maximum ripening sunlight both directly and by reflection from the river surface. For the same reason, all the best Mosel sites face south (which is why the best vineyards may be on either side of this meandering river). The site should also be sheltered from wind and its ripest grapes are likely to come from vines neither so close to the river that morning mist slows ripening, nor above about 200 m / 600 ft. The easily warmed slate soils typical of the region can also help late season ripening. The result is wines unique in the world for their combination of low alcohol (often only about 8 per cent), striking aroma, high extract, and delicacy of texture. No other variety planted here can achieve as much subtlety.

The Pfalz region, on the other hand, the second most important for German Riesling, increased its plantings of Riesling dramatically during the 1980s, to 4,300 ha / 10,600 acres in 1990 from 3,000 ha in 1980. Like neighbouring Rheinhessen, this has always been a region with a particularly varied palette of vine varieties, but the gentle climate of the Mittel Haardt provides Riesling with such shelter and favourable exposition that in many years here it can ripen naturally to produce full bodied dry wines of really spicy, exuberant character, Spätlese Trocken in particular, yet still with a sufficiently extended growing season to keep both acidity and subtlety appetizingly high.

By 1990 the Württemberg region had (just) overtaken the Rheingau as third most important grower of German Riesling, even if most of it is for local consumption and relatively dry and full bodied. The Rheingau, which also had just over 2,500 ha of Riesling planted in 1990, is regarded as Riesling's traditional home and indeed the variety represents 80 per cent of all vine plantings. The best Rheingau Rieslings, increasingly made dry to be drunk with food, represent a faithful statement of their exact provenance. The region is also famous as the original source of Germany's botrytized sweet wines.

Riesling is also the most planted variety in Germany's smaller Mittelrhein and Hessische Bergstrasse regions and has just reasserted itself in terms of area planted over Müller-Thurgau in the Nahe, where the upper reaches of this river yield the finest, most crackling wines. There are another 1,000 ha or so of Riesling in Baden, where the warmer soils rarely show the variety at its best and tend to favour the various PINOT varieties.

For many wine drinkers, Riesling is acceptable only in its French form, a wine from Alsace, the only part of France where this German vine is officially allowed (although several producers in such diverse appellations as Pouilly-Fumé and Barsac surreptitiously rear a row or two for fun). Alsace's plantings of Riesling, regarded as Alsace's noblest variety, have been increasing steadily and passed the 3,000 ha mark in the late 1980s. The great majority of Alsace Riesling is planted in the higher, finer vineyards of the Haut Rhin, where the heavier GEWÜRZTRAMINER covers even more ground. On the flatter land of the Bas Rhin, the soil and climate is not unlike the less interesting parts of the German Pfalz regions and the wines that result can be thin and uninteresting. What is needed to produce Alsace Riesling of real class is, as in Germany, a favoured site of real interest such as many of Alsace's famous grands crus vineyards.

The hallmark of Alsace has been dry wines from aromatic grapes such as Riesling and certainly the great majority of Alsace Rieslings follow the variety's alluring perfume with a taste that is fairly alcoholic (easily 12 per cent) and bone dry. The dry climate of Alsace minimizes the risk of rot and makes extended ripening a real possibility, however, often resulting in the prized late harvest wines of varying degrees of sweetness.

To the north, about 10 per cent of the Luxembourg vineyard is planted with Riesling, which tends to produce dry, relatively full bodied wines closer to Alsace in style than to the Mosel-Saar-Ruwer, which is just over the German border.

In Austria Riesling, sometimes called Rheinriesling or Weisser Riesling to distinguish it from the more widely planted WELSCHRIESLING, is qualitatively unimportant, covering just over 1,000 ha/2,500 acres of vineyard, but is regarded as one of the country's finest wines when made on a favoured site. The most hallowed Austrian Rieslings are dry, whistle-clean, concentrated and aromatic, and come from the terraced vineyards of the Wachau and neighbouring Kremstal and Kamptal in Lower Austria. Good Riesling is also made in the vineyards round Vienna.

Not surprisingly, Riesling works well in the continental climate of Slovakia to the immediate north of Austria's vineyards, where relatively light wines have real crackle and race. Most of Switzerland is too cool to ripen Riesling properly, with the exception of some of the more schistous soils and warmest vineyards of the Valais around Sion.

Although practically unknown in Iberia, Riesling has infiltrated the far north east of Italy. It is grown with real enthusiasm in the high vineyards of Alto Adige, where it produces delicate aromatic wines quite unlike most Italian whites. It is also grown quite successfully in Friuli, where it is known as **Riesling Renano**, and just over the border in Slovenia where it may be called Rheinriesling. Riesling, known as Rizling Rajinski and variants thereof, is also planted southwards through what was Yugoslavia in Croatia and, less distinctively, in Vojvodina.

It is planted throughout the rest of eastern Europe in Hungary and Bulgaria and to a much more limited extent in Romania, but in each of these countries the climate can be too warm to coax much excitement from the variety and Welschriesling tends to predominate.

The ex-Soviet Union may have even more Riesling planted than Germany. The last available official statistics, from the mid 1980s, suggest that there were 25,000 ha/62,000 acres of true Riesling before Gorbachev's vine pull scheme. It is easy to see why the variety would be popular in the cold winters of Russia, and Ukraine which has by far the biggest area planted with the variety. Rhine Riesling is also grown in Moldova and in most of the Central Asian republics.

In the New World true Riesling is most widely grown in Australia, where it was the most planted white wine grape variety until Chardonnay caught up with its nearly 4,000 ha/10,000 acres in 1990. Here, known as Rhine Riesling, it has received little of the respect it deserves, precisely because of its ubiquity (and perhaps also because the Australians had at one time a confusing tendency to call just about any white grape variety Riesling). It is associated most intimately with South Australia, where it can yield fine, dense lime-scented wines in the cooler reaches of the Clare Valley, and more floral examples in the high Eden Valley. These wines age beautifully, sometimes acquiring toastiness, and some late harvest styles have also been made successfully.

New Zealand began to produce convincing wines from its

nearly 300 ha/750 acres of Riesling in the late 1980s, notably when some producers addressed themselves to making scintillating late harvest sweet wines.

Riesling is cultivated far more widely in South America than one might think wise. Argentina has about 1,300 ha, mainly in the hot, irrigated vineyards of Mendoza province. Chile has a little (Santa Monica winery succeeds with it), and there are plantings all over the rest of the continent, often, although not exclusively, in vineyards which ripen far too fast.

Riesling's progress in North America has been hampered simply by consumer demand for anything *but* Riesling. California's total acreage of what is known there as Johannisberg Riesling or White Riesling remained at around 4,000 acres/1,600 ha throughout the 1980s (while Chardonnay plantings grew from 22,000 to 56,000 acres in the decade to 1992). The variety is rarely made bone dry in California, and can reliably command a decent price only if very sweet and described as Select Late Harvest or somesuch.

Washington State claims a special affinity for Riesling, even organizing the world's first truly international conference on the subject, although total area planted has been declining and was only just over 2,000 acres in 1991. As in Oregon, the variety suffers from consumer passion for other varieties rather than from any inherent viticultural disadavantage, indeed some Washington Rieslings can be delightfully delicate.

Because of its winter hardiness, Riesling tends to be treasured in the coolest wine regions of North America. In Canada Riesling is cultivated with particular success in Ontario, just over the border from the Finger Lakes region of New York State, where it is also treasured, not least for its ability to yield commercially interesting ice wine. It was overtaken by CHARDONNAY as most planted *vinifera* vine in Ontario as recently as 1993. A little remains in British Columbia.

RIESLING ITALICO, or **RIESLING ITALIANSKI,** synonyms for WELSCHRIESLING, although the Germans would like to see the word Riesling replaced by 'Rizling' unless the vine in question is their own great RIESLING.

RIESLING-SYLVANER, deceptively enticing name for MÜLLER-THURGAU, used in Switzerland and New Zealand.

RIVANER, Luxembourg name for MÜLLER-THURGAU, sometimes used in Slovenia.

RIZLING, term used in many of the varied names for WELSCHRIESLING, notably Olasz Rizling and Laski Rizling used respectively in Hungary and Slovenia. The variety is known as **Rizling Vlassky** in parts of what was Czechoslovakia.

🍇 **RKATSITELI,** Russian grape planted much more extensively than most western wine drinkers realize. As the most planted grape in the ex-Soviet Union, it was the world's third most important wine vine (after AIRÉN and GRENACHE) until President Gorbachev's widespread uprooting of vineyards in the mid 1980s. Rkatsiteli, being grown in all Soviet wine-producing republics with the exception of Turkmenistan, was a chief casualty of this policy but is still one of the world's four most planted white wine grapes, on vineyards probably totalling 262,000 ha/647,000 acres in the CIS alone in the early 1990s. It is particularly important in Georgia where Rkatsiteli, with 52,000 ha/128,500 acres, is the most planted grape, and it is the second most planted grape in both Moldova and the Ukraine. It is well adapted to particularly cold winters, which makes it popular in Russia too. Much is demanded of this variety and it achieves much, providing a base for a wide range of wine styles, including fortified wines and brandy. The wine is distinguished by a keen level of acidity and by good sugar levels.

The variety is also widely cultivated in Bulgaria, where it has been the country's most important white grape variety by far, planted on about 22,000 ha/55,000 acres in 1993—and it is also grown in Romania. There are even pockets of Rkatsiteli in the United States.

In China it is known as Baiyu and has been an important source of neutral white wine for the nascent Chinese wine industry.

🍷 **ROBOLA,** speciality of the Ionian island of Cephalonia in Greece. The distinctively powerful, lemony dry varietal version white is one of the most refined Greek whites. The vine is almost certainly the same as Rebula of Slovenia and RIBOLLA, which has been grown in Friuli in north east Italy since the 13th century and was probably brought to Cephalonia by Venetian merchants. The wine made from these early ripening grapes is high in both acidity and extract and is much prized within Greece.

🍷 **RODITIS, RHODITIS,** slightly pink-skinned Greek grape variety traditionally grown in the Peloponnese, especially before phylloxera struck Greek vineyards. The vine is particularly sensitive to powdery mildew. It ripens relatively late and keeps its acidity quite well even in such hot climates as that of Ankialos in Thessaly in central Greece, although it can also ripen well, and makes very much more interesting wine, in high-altitude vineyards. It is often blended with the softer SAVATIANO, particularly for retsina.

🍷 **ROLLE,** Provençal vine originally associated with Bellet near Nice but now increasingly grown in the Languedoc and, especially, Roussillon. It is aromatic and usefully crisp for warm wine regions and is accepted by French authorities as identical to the VERMENTINO of Corsica, Sardinia, and southern Italy. Some Italian authorities maintain it is not identical to the **Rollo** of Liguria.

ROMAN MUSCAT, common name for MUSCAT OF ALEXANDRIA,

thought to have been particularly widely dispersed around the Mediterranean by the Romans.

🍇 **ROMORANTIN**, eastern Loire grape fast fading from the French *vignoble*. Cour Cheverny is an appellation especially created for Romorantin grown just west of Blois.

🍷 **RONDINELLA**, Italian red grape variety grown in the Veneto, especially for Valpolicella. The vine yields profusely and is therefore extremely popular with growers but its produce is rarely sufficiently flavoursome to please consumers. Not, fortunately, as widely planted as CORVINA with a total of just over 2,800 ha/7,000 acres in 1990.

RORIZ, or **Tinta Roriz,** most common of several Portuguese names for the Spanish red wine grape variety TEMPRANILLO, used particularly in the Douro Valley, where it is one of the most planted varieties for the production of port. Others include Aragonez and Tinto de Santiago.

ROSENMUSKATELLER, name for pink-berried form of MUSCAT BLANC À PETITS GRAINS in the Italian Tyrol.

🍷 **ROSETTE**, or Seibel 1000, old French hybrid once grown in New York State, where it produced pale wines.

ROSIOARA. See PAMID.

🍷 **ROSSESE**, esteemed grape producing distinctively flavoured varietal wines in the north west Italian region of Liguria. The variety has a long history in the region and it has its own DOC in the west of the region in Dolceacqua, whose wines are admired, though variable.

🍷 **ROSSIGNOLA**, optional, tart ingredient in Valpolicella.

🍇 **ROTGIPFLER**, marginally less noble of the two grapes traditionally associated with the spicy, full wines of Gumpoldskirchen south of Vienna. (The other is ZIERFANDLER.) At the end of the 1980s there were about 200 ha/500 acres of Rotgipfler, which ripens late, but earlier than Zierfandler, and whose wines are particularly high in extract, alcohol, and bouquet.

ROUCHALIN, **ROUCHELIN**, occasional name's for CHENIN BLANC in South West France.

ROUCHET, Italian red grape variety. See RUCHÈ.

🍇 **ROUPEIRO**, Portuguese variety grown particularly in the Alentejo producing basic white wine to be drunk as young as possible. It is known as Códega in the Douro and sometimes as Alva.

🍇 **ROUSSANNE**, doubtless named after the russet or *roux* colour of its grapes, and one of only two grapes allowed in the white

159

versions of the northern Rhône's classic red wine appellations Hermitage, Crozes-Hermitage, and St-Joseph and into the exclusively white but often sparkling St-Péray. In each of these appellations MARSANNE, its traditional blending partner here, is far more widely grown because, although the wine produced is not as fine, there is more of it. Roussanne's irregular yields, tendency to powdery mildew and rot, and poor wind resistance all but eradicated it until better clones were selected. Even today it is preferred by a minority of producers such as Jaboulet—although it is increasingly being planted to add interest to the whites of the southern Rhône, Languedoc, and Roussillon, where Roussanne's tendency to ripen late is less problematic than in the north.

Roussanne's chief attribute is its haunting aroma, something akin to a particularly refreshing herb tea, together with acidity that allows it to age much more gracefully than Marsanne which, in blends, can lend useful body. It does need to reach full maturity, however, in order to express itself elegantly. Roussanne (but not Marsanne) is one of four grape varieties allowed into white Châteauneuf-du-Pape and Château de Beaucastel here has demonstrated that carefully grown Roussanne can respond well to oak ageing. The variety is also grown in Provence (although the more common pink-berried **Roussanne du Var** is a lesser, unrelated variety used for vins de pays and ordinary table wines). Although it is usually classified with Marsanne and VERMENTINO in appellation regulations, it can also make a fine blending partner with the fuller bodied CHARDONNAY. It can suffer in drought conditions.

The variety is also beguilingly fine and aromatic at Chignin in Savoie, where it is known as Bergeron, but should not be confused with ROUSSETTE. It is grown to a limited extent in Liguria and Tuscany, where it is a permitted ingredient in Italy's Montecarlo Bianco, and can also be found in Australia, presumably having been taken there as a North Rhône partner to the much more successful SHIRAZ.

🍇 ROUSSETTE, Savoie's most exciting white grape, also known as Altesse (a name signifying a certain loftiness of reputation). Its origins are rich in mystery and intrigue. The widespread influence of the House of Savoie in the Middle Ages was such that this corner of France has its own distinctive vine varieties with possible links beyond what is now the French border (see MONDEUSE, TROUSSEAU, SAVAGNIN, POULSARD). Altesse was for long thought to have been imported from Cyprus, but Galet reports that, grown alongside the famous FURMINT of Hungary, the Savoie variety is virtually indistinguishable. The variety is a shy, late producer but it resists rot well and the wine produced is (like Furmint) relatively exotically perfumed, has good acidity, and is well worth ageing. In recognition of Roussette's superiority over the more common JACQUÈRE, Roussette de Savoie has its own appellation. It is also grown in Bugey. Also, a Rhône synonym for ROUSSANNE.

🍇 **ROYALTY**, also known as **ROYALTY 1390**, red-fleshed hybrid bred in California by crossing the progenitor of all red-fleshed TEINTURIERS, Alicante Ganzin, with the Jura grape variety of TROUSSEAU and released in 1958 along with the somewhat similar but much more successful RUBIRED. There is little regal about this particular variety which is difficult to grow, although more than 800 acres/320 ha persist, almost exclusively in the Central Valley.

🍇 **RUBIRED**, popular red-fleshed California hybrid, the result of crossing the progenitor of all red-fleshed TEINTURIERS, Alicante Ganzin, with the port variety TINTO CÃO. It was released with the much less successful ROYALTY in 1958. Its productivity and depth of colour has made the variety popular with blenders of wine, California port, and grape juice so that there were still more than 6,000 acres/2,500 ha in the Central Valley in the early 1990s. Unlike Royalty, it is easy to grow and has been tested in Australia.

🍇 **RUBY CABERNET**, once popular California crossing bred in 1949 by Dr H. P. Olmo of the University of California at Davis (see also EMERALD RIESLING, CARNELIAN). CARIGNAN was crossed with CABERNET SAUVIGNON in an attempt to combine Cabernet characteristics with Carignan productivity and heat tolerance. Ruby Cabernet had its heyday in the 1960s. Although it has faded as a varietal wine since then, it may deserve a better fate. Even though it was designed to yield claret-like wines from hot regions, it has done surprisingly well in cooler areas. Total California plantings have remained static since the 1970s at just under 7,000 acres/2,800 ha, most notably in the southern central San Joaquin Valley. It is also grown by several producers in South Africa, where it occasionally yields as much as five times as heavily as Cabernet Sauvignon in the same conditions, and has also been grown to a very limited exent in Australia.

🍇 **RUCHÈ, ROUCHET**, or occasionally **ROCHE**, relatively obscure Piedmont vine, enjoying something of a revival with its own varietal DOC around Castagnole Monferrato, occasionally labelled Rouchet. Like NEBBIOLO, the wine is headily scented and its tannins imbue it with an almost bitter after-taste. According to Gleave, it is locally supposed to have been brought from Burgundy in the 18th century.

🍇 **RUFETE**, early ripening port vine making lightish wine which can oxidize easily in the greater Douro basin both in Spain and Portugal. According to Mayson it may the same as the Dão region's TINTA PINHEIRA.

RUFFIAC, also known as ARRUFIAC.

RULÄNDER

⚘ RULÄNDER, main German name for PINOT GRIS, which was propagated in the Pfalz in the early years of the 18th century by wine merchant Johann Seeger Ruland. Since the mid 1980s some German producers have used this synonym to differentiate sweeter styles, often made from nobly rotten grapes, from a drier wine that would be labelled Grauburgunder. Ruländer needs a good site with deep, heavy soils to maximize the impressive level of extract of which it is capable. It can easily build up far higher grape sugars than the classic RIESLING on the same sort of site although it is rarely in direct competition for territory. It is planted in most of the wine regions of Germany and is quite important in Eastern Germany and the south, although it is relatively rare in the Mosel. More than half of Germany's total Ruländer plantings are in the relatively warm Baden region, although there are several hundred ha in both Rheinhessen and Pfalz. Because of its inherent fatness, it can be one of Germany's more successful varieties when styled as a dry wine.

In most German-speaking wine regions—Germany, Austria, and Italy's Alto Adige—Pinot Gris is regarded as rather more ordinary than Pinot Blanc, although it can quite easily reach higher ripeness levels. In Austria, where its wines are typically even earthier and richer than their German counterparts, it is also commonly called Ruländer and is also much less common than Pinot Blanc, covering just 1 per cent of the country's vineyard, chiefly in Styria and Burgenland.

The name Ruländer is also used for some of the Pinot Gris that is widely planted in Romania.

RUSA, occasional Romanian name for GEWÜRZTRAMINER.

RUZICA, alternative name for DINKA.

S

⚘ SACY, grown particularly in the Yonne *département* near Chablis country. At one time it threatened to usurp all manner of nobler vines in eastern France, most notably CHARDONNAY, from southern Burgundy in the 18th century and from the vineyards of Chablis in the early 20th century. Its productivity is the vine's chief attribute, its acidity the wine's most noticeable characteristic. This has been used to reasonable effect by sparkling wine producers and the variety, called here Tresallier, is still an important ingredient in the white wines of St-Pourçain.

⚘ SAGRANTINO, lively, sometimes tannic red grape grown in Umbria in central Italy, particularly in the Montefalco area. Sagrantino di Montefalco was elevated to DOCG status in the mid 1990s. Sagrantino has been used as an ingredient in dried grape wines but today shows promise as a carefully vinified dry red, sometimes blended with SANGIOVESE.

ST-ÉMILION, name used for UGNI BLANC in the Cognac region.

ST GEORGE, Anglicized name for the Greek variety AGIORGITIKO.

🍇 ST LAURENT, quantitatively and qualitatively important eastern European grape, particularly in Lower Austria. It shares many of PINOT NOIR'S characteristics (juicy fruit and relatively low natural grape tannins) and is thought by some to be related to the great red burgundy grape. It ripens well ahead of Pinot Noir, however, and can be cultivated on a much wider range of sites. Its thicker grape skins also help ward off rot, although coulure can be a problem. Its only major viticultural disadvantage is dangerously early budding, although this is no great problem in Burgenland in the warm south eastern corner of Austria, where it is capable of producing deep-coloured, velvety reds with sufficient concentration—provided yields are limited—to merit ageing in oak and then bottle. Lesser versions can be simply and soupily sweet but the variety has been successfully blended with such fashionable varieties as CABERNET SAUVIGNON, Blauburgunder (Pinot Noir), and also with Austria's livelier BLAUFRÄNKISCH, notably on the shores of the Neusiedlersee.

It has been known in eastern France and in the German Pfalz region, but is grown in much greater quantity in Slovakia, and the Czech Republic where it is also known as Vavrinecke, or Svatovavrinecke.

🍇 ST-MACAIRE, unidentified California grape producing rather ordinary table wine.

🍇 STE MARIE, Gascon rarity.

🍇 ST-PIERRE DORÉ, almost extinct, very productive central French vine found occasionally in St-Pourçain.

🍇 SALVADOR, California red-fleshed hybrid vine superseded by RUBIRED and therefore in decline.

SALVAGNIN (NOIR) is a Jura name for PINOT NOIR, disconcertingly similar to the name of one of the Jura's own vine varieties, SAVAGNIN.

SÄMLING 88, common Austrian synonym for the SCHEUREBE grape which has not shone there, although several hundred ha are planted in southern Austrian wine regions of Burgenland and Styria.

SAMSÓ, Penedès name for CARIGNAN.

SAMTROT. See MÜLLERREBE.

🍇 SANGIOVESE, extremely variable red grape that is Italy's most planted and is particularly common in central Italy. In 1990 almost 10 per cent of all Italian vineyards, or more than 100,000 ha / 250,000 acres, were planted with some form of Sangiovese.

Wines have noticeable tannins and acidity, not always a great depth of colour but a character that varies from farmyard/leather to plum/prune according to quality and ripeness.

Like PINOT, Sangiovese mutates easily. In its various clonal variations and names (**Sangioveto**, Brunello, Prugnolo Gentile, Morellino, and many, many more) Sangiovese is the principal vine variety for fine red wine in Tuscany, the sole grape permitted for Brunello di Montalcino and the base of the blend for Chianti, Vino Nobile di Montepulciano, and the vast majority of Supertuscans (highly priced sophisticated blends in the main). Widely planted in Umbria, it gives its best results in Torgiano and Montefalco. In the Marches it is the base of Rosso Piceno and an important component of Rosso Conero. Sangiovese is also planted in Latium and can be found as far afield as Lombardy and Valpolicella to the north and Campania to the south.

The vine itself, probably indigenous to Tuscany, is of ancient origin, as the literal translation of its name ('blood of Jove') suggests, and it has been postulated that it was even known to the Etruscans—although the first historical mention of the variety dates only from the early 18th century. It was the major ingredient in Baron Ricasoli's 19th century recipe for Chianti, although its natural hardness was softened by blending with CANAIOLO. MAMMOLO and COLORINO as well as the white grapes MALVASIA and, especially, TREBBIANO were subsequent ingredients.

Conventional ampelographical descriptions of Sangiovese, based on the pioneering work of Molon in 1906, divide the variety into two families: the superior **Sangiovese Grosso**, to which Brunello, Prugnolo Gentile, and the Sangiovese di Lamole (of Greve in Chianti) belong, and the **Sangiovese Piccolo** of other zones of Tuscany. This classification may be too simplistic, however, for there are so many different clones that no specific qualitative judgements can be based on either the size of the berries or the bunches. Significant efforts are at last being made to identify and propagate superior clones. The variety adapts well to a wide variety of soils, although the presence of limestone seems to exalt the elegant and forceful aromas that are perhaps the most attractive quality of the grape.

Sangiovese ripens late, which gives rich, alcoholic, and long-lived wine in hot years and creates problems of high acidity and hard tannins in cool years. Over-production tends to accentuate the wine's acidity and lighten its colour, which can also start to brown relatively early. Thin grape skins can lead to rot in cool and damp years, which is a serious disadvantage in a region where autumn rains are commonplace. Too often Sangiovese has been planted with scant attention to exposure and altitude in Tuscany's high vineyards.

Throughout modern Tuscany, Sangiovese is now often blended with a certain proportion of CABERNET SAUVIGNON, whether for Chianti (in which case the interloper should not exceed 10 per cent of the total) or a Supertuscan. This highly successful blend, in which the intense fruit and colour of Cabernet marries well with the characterful native variety, was first sanctioned by the authorities in the wines of Carmignano.

But in terms of quantity rather than quality, Sangiovese is most important in Romagna north and east of Tuscany, where Sangiovese di Romagna is as common as the LAMBRUSCO vine is in Emilia. Sangiovese di Romagna wine is typically light, red, ubiquitous, and destined, quite properly, for early consumption. The most widely planted Sangiovese vines planted in Romagna appear to have little in common with Tuscany's most revered selections, although there are signs that this situation is improving. Some Sangiovese is grown in the south of Italy, where it is usually used for blending with local grapes, and the success of Supertuscans has inevitably led to a certain amount of experimentation with the variety to the north of Tuscany too.

Corsica's NIELLUCCIO is none other than Sangiovese.

Like other Italian grape varieties, particularly red ones, Sangiovese was taken west, to both North and South America, by Italian emigrants. In South America it is best known in Argentina, where there are several thousand ha, mainly in Mendoza province, producing wine that few Tuscan tasters would recognize as Sangiovese.

In California, however, international recognition for the quality of Supertuscans (and a fashion for all things Italian) brought a sudden increase in Sangiovese's popularity in the late 1980s and early 1990s. In 1991 its acreage was climbing towards 200 (80 ha), or about as much land as Cabernet Sauvignon had commanded in 1961. Most substantial plantings are in the Napa Valley, but smaller patches can be found in Sonoma County, San Luis Obispo, and the Sierra Foothills. Early results are extremely varied but patently more successful than California versions of the great NEBBIOLO of Piedmont.

SANTARÉM, Ribatejo synonym for the Portuguese CASTELÃO FRANCÊS variety.

♥ SAPERAVI, Russian grape variety notable for the colour and acidity it can bring to a blend. As a varietal wine, it is capable, not to say demanding, of long bottle ageing. The flesh of this dark-skinned grape is deep pink. Saperavi ripens late, is relatively productive, and is quite well adapted to the cold Russian winters, but not so well that the Russian Potapenko viticultural research institute was discouraged from producing a **Saperavi Severny**, a hybrid of SEVERNY and Saperavi which was released in 1947 and incorporates not just Saperavi's *vinifera* genes, but also those of the cold-hardy *Vitis amurensis*.

Traditional Saperavi is planted throughout almost all of the wine regions of the ex-Soviet Union. It is an important variety in Russia, Ukraine, Moldova, Georgia, Kazakhstan, Uzbekistan, Tajikistan, Kyrghyzstan, and Turkmenistan, although in cooler areas the acidity may be too marked for any purpose other than blending, despite its relatively high sugar levels. It has also been grown in Bulgaria for some time.

At Magaratch, the Crimean wine research centre, Cabernet Sauvignon and Saperavi have been crossed to produce the promising MAGARATCH RUBY.

🍇 **SÁRFEHÉR**, undistinguished, high-yielding Hungarian variety traditionally grown on the sandy Great Plain for table grapes and sparkling wines.

SARGAMUSKOTALY, occasional Hungarian name for MUSCAT BLANC À PETITS GRAINS.

🍇 **SAUVIGNON BLANC**, piercingly aromatic, crisp variety solely responsible for some of the world's most popular, and most distinctive, dry white wines: Sancerre, Pouilly-Fumé, and a host of Sauvignons and Fumé Blancs from outside France. In many great white wines both dry and sweet it also adds nerve and zest to its most common blending partner SÉMILLON. Like the famous and quite distinct black-berried vine CABERNET SAUVIGNON, Sauvignon Blanc seems to have its origins in Bordeaux, where it has been enjoying a revival in popularity.

The variety is often simply called **Sauvignon**, especially on wine labels, but there are **Sauvignons Jaune, Noir, Rose,** and **Violet,** according to the colour of the berries. **Sauvignon Gris** is another name for Sauvignon Rose and has discernibly pink skins. It can produce more substantial wines than many a Sauvignon Blanc, and has a certain following in Bordeaux and the Loire. See also FIÉ and SAUVIGNON VERT.

Sauvignon Blanc's most recognizable characteristic is its piercing, instantly recognizable aroma. Descriptions typically include 'grassy, herbaceous, musky, green fruits' (especially gooseberries), 'nettles', and even 'tomcats'. Scientists tell us that the aroma compounds responsible are methoxypyrazines. Over-productive Sauvignon vines planted on heavy soils can produce wines only vaguely suggestive of this but Sauvignon cautiously cultivated in the central vineyards of the Loire, unmasked by oak, can reach the dry white apogee of Sauvignon fruit with some of the purest, most refreshingly zesty wines in the world. The best Sancerres and Pouilly-Fumés served as a model for early exponents of New World Sauvignon Blanc, although by the 1980s it was the Loire vignerons who copied their counterparts in California, Australia, and New Zealand (which achieved rapid fame with this variety) in experimenting with fermentation and maturation in oak. Oak-aged examples usually need an additional year or two to show their best, but almost all dry, unblended Sauvignon is designed to be drunk young, although there are both Loire and Bordeaux examples that can demonstrate durability, if rarely evolution, with up to 15 years in bottle. As an ingredient with Sémillon and sometimes MUSCADELLE in the great sweet white wines of Sauternes, on the other hand, Sauvignon plays a minor but important part in one of the world's longest-living wines.

The vine is particularly vigorous, which has caused problems in parts of the Loire and New Zealand. If the Sauvignon vine's vegetation gets out of hand, Sauvignon grapes fail to reach full maturity and the resulting wine can be aggressively herbaceous, almost intrusively rank. (Underripe Sémillon can exhibit very similar characteristics—just as underripe Cabernet Sauvignon can smell like CABERNET FRANC.) A low-vigour rootstock and canopy management can help combat this problem.

Sauvignon buds after but flowers before Sémillon, with which it is typically blended in Bordeaux and, increasingly, elsewhere. Until suitable clones such as 297 and 316 were identified, and sprays to combat Sauvignon's susceptibility to fungal diseases developed, yields were uneconomically irregular. In 1968, for example, Sauvignon was France's 13th most planted white grape variety, but within 20 years had risen to fourth place.

With its common blending partner Sémillon, Sauvignon is planted all over South West France, particularly in Bergerac. In Bordeaux Sémillon is still much more commonly planted than Sauvignon Blanc, which is concentrated in the Entre-Deux-Mers, Graves, and the sweet wine-producing districts in and around Sauternes. It is perhaps no coincidence that the average Loire Sauvignon has more Sauvignon character than the average all-Sauvignon Bordeaux Blanc when the official maximum yield for the first is 10 hl/ha lower than the 65 hl/ha (3.7 tons/acres) allowed in Bordeaux.

It is in the Loire that Sauvignon is encountered in its purest, most unadulterated form. In the often limestone vineyards of Sancerre, Pouilly-sur-Loire (whose Sauvignons are said to be *fumé*, or 'smoked' by the local flinty soil) and their eastern satellites Quincy, Reuilly, and Menetou-Salon it can demonstrate one of the most eloquent arguments for marrying variety with suitable terroir. The variety is often called Blanc Fumé here and has happily replaced most of the lesser varieties once common, notably much of the CHASSELAS in Pouilly-sur-Loire. Most of these wines are designed to be drunk, well chilled, within two years and are none the worse for that.

From this concentration of vineyards, which might well be considered the Sauvignon capital of the world (a title disputed by the inhabitants of Marlborough in New Zealand's South Island), Sauvignon's influence radiates outwards: north east towards Chablis in Sauvignon de St-Bris, south to St-Pourçain-sur-Sioule, and north and west to Coteaux du Giennois, and Cheverny as well as to a substantial quantity of eastern Loire wines, typically labelled Touraine. Such Sauvignons tend to be light, racy, and, of course, aromatic. With Chardonnay, it has also been allowed into the vineyards of Anjou where it is often blended with the indigenous CHENIN BLANC.

Elsewhere in France Sauvignon Blanc has been an obvious, though not invariably successful, choice for those seeking to make internationally saleable wine in the Languedoc (yields have often been too high to extract sufficiently varietal character from the vine) and small plantings of Sauvignon can be found in some of the Provençal appellations.

Across the alps, Sauvignon's most successful Italian region is the far north east in Friuli with some Alto Adige and Collio examples exhibiting extremely fine fruit and purity of flavour. In the 1980s, Italy's plantings of Sauvignon doubled to nearly 3,000 ha/7,400 acres. Slovenia and especially Styria in south east Austria also have a masterly hand with this variety, combining fruit with aroma. As 'Muskat-Silvaner' (as the variety is known in German) it is grown but rarely in Germany, where many would argue that young Riesling can provide the same sort of crisp, aromatic white. It is planted to a certain extent further south—even if the wines tend to be progressively heavier and sweeter. Parts of Serbia, the Fruš ka Gora district of Vojvodina, and some of Slovakia clearly has potential. Romania had nearly 5,000 ha/12,000 acres of Sauvignon Blanc in the early 1990s. Of the ex-Soviet republics, both Ukraine and, especially, Moldova have sizeable plantings of the variety.

Sauvignon Blanc has been imported into Iberia by only the most dedicated internationalists and certainly Portugal and north western Spain have no shortage of indigenous varieties capable of reproducing similar wine styles. There is a tendency for Sauvignon Blanc to taste oily when reared in too warm a climate, as it sometimes does in Israel and other Mediterranean vineyards where those with an eye to the export market put it through its paces.

This was clearly perceptible in many of Australia's early attempts with the variety, although by the early 1990s there was even keener appreciation of the need to reserve it for the country's cooler sites such as the Adelaide Hills. Chardonnay plantings outnumber those of Sauvignon Blanc by more than four to one in Australia.

In New Zealand on the other hand, only Müller-Thurgau and Chardonnay cover more ground, and this relatively minute wine industry can boast almost as big an area planted with Sauvignon (800 ha/2,000 acres by 1992) as Australia. This is the variety that introduced New Zealand wine to the world and did it by developing its own style: intensely perfumed, pungent, more obviously fruity than the Loire prototype, and with just a hint of both gas and sweetness. This style of Sauvignon can now be found in South and the cooler areas of North America, in the south of France and, doubtless, even further afield before long.

'Sauvignon' is the white wine most commonly exported from Chile, but according to official statistics in the early 1990s, Sauvignon Blanc accounted for less than 5 per cent of the country's total wine production (while Semillon accounted for more than 26 per cent). Much of Chile's 'Sauvignon' is SAUVIGNON VERT, although there are strenuous efforts to replace this with true

Sauvignon Blanc, which has shown a real affinity with the ocean-cooled Casablanca Valley. California clones of Sauvignon Blanc have been widely planted in Chile but tend to suffer from excessive vigour. High yields also help depress the keynote aromas of Sauvignon Blanc in other South American wines labelled Sauvignon, including those made from Argentina's 600 ha / 1,500 acres of Sauvignon (as opposed to 800 ha of TOCAI FRIULANO and 2,000 ha of Semillon counted in 1989), although some attention is being focused on Sauvignon within Argentine companies keen to export. Brazil's 'Sauvignon' is usually SEYVAL, according to Galet.

Thanks to Robert Mondavi, California's most cosmopolitan wine producer, who renamed it Fumé Blanc, Sauvignon Blanc enjoyed enormous success in California in the 1980s and the state's total plantings were more than 12,000 acres / 4,800 ha in 1994, one-third of it in Napa, where problems of vine vigour were largely overcome by the late 1980s. Some of the wines are sweet and even botrytized, a sort of Semillon-free Sauternes. There has also been an increase, as elsewhere in the New World, in blending in some Semillon to dry white Sauvignon to add weight and fruit to Sauvignon's aroma and acidity. Like California, Washington State makes both Sauvignon Blanc and Fumé Blanc from its 900 acres (in 1994) of Sauvignon, a poor third to Chardonnay and Riesling. Of other American states, Texas has had particular success with the variety.

But perhaps Sauvignon's real success has been in South Africa where, for want of genuine Chardonnay perhaps (see AUXERROIS) local wine drinkers fell upon the Cape's more successful early Sauvignons as a fashionable internationally recognized wine style. In 1994 there were nearly 4,000 ha / 10,000 acres of Sauvignon Blanc in South Africa, capable of their own slightly softer, more marine twist to the standard New World recipe. South Africans regularly blend Sauvignon Blanc with CHARDONNAY (a blend that can work surprisingly well) and other desirable and not so desirable varieties.

🍇 **SAUVIGNON VERT,** also known as **Sauvignonasse**, quite distinct from and more rustic than the more famous SAUVIGNON BLANC. This is the Sauvignon most commonly planted in Chile and can produce wines that taste like high-yielding Sauvignon Blanc but without much extract or ageing potential. Galet maintains that Sauvignon Vert is identical to the TOCAI FRIULANO of north east Italy—a theory not embraced by all Italians. The Montpellier ampelographer also maintains that the variety called Sauvignon Vert in California is in fact the MUSCADELLE of Bordeaux.

🍇 **SAVAGNIN,** fine but curious vine variety with small, round, white berries. In the Jura region of France it is as much a viticultural curiosity as the wine it alone produces, *vin jaune*, is a wine-making oddity. Today only those producers rewarded by the high prices fetched by *vin jaune* would presumably persist with a vine that can yield so churlishly.

It may be included in any Jura white wine but is usually in practice reserved for the Jura's extraordinary, sherry-like *vin jaune*. It is well adapted to the ancient, west-facing slopes of Jura but many believe it is at its finest in what remains of the vineyards of Château-Chalon, where it may sometimes be left to ripen as late as December.

Called Gringet, it is also a minor ingredient in the sparkling wines of Ayse in Savoie.

Galet maintains that Savagnin is identical to the TRAMINER which was once grown widely in Germany, Alsace, Hungary, and Austria, and that GEWÜRZTRAMINER is the pink-berried MUSQUÉ mutation of Savagnin. Certainly Austrian Traminer is, like Jura's Savagnin, famous for its aroma and ability to age, and a non-musqué Savagnin Rosé is still cultivated to a very limited extent in Alsace, where it is sometimes called KLEVNER d'Heiligenstein.

SAVAGNIN NOIR is a Jura name for PINOT NOIR.

🍇 **SAVATIANO,** Greece's most common wine grape, widely planted, on up to 20,000 ha / 50,000 acres throughout Attica and central Greece. This vine, with its exceptionally good drought resistance, is the most common ingredient in retsina, although RODITIS and ASSYRTICO are sometimes added to compensate for Savatiano's naturally low acidity. On particularly suitable sites, picked relatively early and vinified carefully, Savatiano can produce well-balanced dry white wines.

🍇 **SCHEUREBE,** often called simply **Scheu**, the one early 20th century German crossing that deserves attention from any connoisseur, and the only one named after the prolific vine breeder Dr Georg Scheu, the original director of the viticultural institute at Alzey in the Rheinhessen. Like KERNER and EHRENFELSER, it can produce some top quality wines. RIESLING genes would seem a prerequisite for quality, and Scheurebe was a SILVANER × Riesling cross, but it is much more than a riper, more productive replica of Riesling. Provided it reaches full maturity (like such other 20th century crossings as BACCHUS and ORTEGA it is distinctly unappetizing if picked too early), Scheurebe wines have their own exuberant, racy flavours of blackcurrants or even rich grapefruit. Some Pfalz examples have been particularly exciting and Scheu is one of the few varietal parvenus countenanced by quality-conscious German wine producers, not just because it ripens so well, but because grape sugars are so delicately counterbalanced with the nerve of acidity—perhaps not quite so much as in an equivalent Riesling, but enough to preserve the wine for many years in bottle.

Scheurebe can be relied upon not only to ripen more dramatically than Riesling, but also to yield considerably more wine, often 100 hl/ha (5.7 tons/acre). It is not, however, as bountiful as Kerner, Bacchus, and the dreaded MORIO-MUSKAT, and was decisively overtaken in terms of total area planted by Kerner in the 1980s although there were still nearly 4,000 ha/10,000 acres of Scheurebe planted in Germany in the early 1990s, mainly in Rheinhessen and, to a lesser extent, in Pfalz. It also needs a relatively good site, often one which could otherwise support the great Riesling, and young Scheurebe vines are prey to frost damage. Ambitious growers can rely on good frost resistance in mature vines and on Scheurebe's useful encouragement of noble rot in good years. The variety can produce extremely fine Beerenauslese and Trockenbeerenauslese which may not last quite as many decades as their Riesling counterparts but are much less rare and therefore better value.

The variety is also grown in southern Austria, where it is known as Sämling 88.

🌿 SCHIAVA, undistinguished north Italian vine known as Vernatsch by the German speakers of the Alto Adige, or Sudtirol as they would call it; and as TROLLINGER in the German region of Württemberg where it is widely grown. The name Schiava, meaning 'slave', is thought by some to indicate Slav origins.

It is most planted in Trentino-Alto Adige in northern Italy where several forms are known. The most common is **Schiava Grossa** (Grossvernatsch), which is extremely productive but is not associated with wines of any real character or concentration. Total Italian plantings of Schiava Grossa declined by about a quarter in the 1980s to 3,400 ha/8,400 acres. **Schiava Gentile** (Kleinvernatsch) produces better-quality, aromatic wines from smaller grapes, and total plantings had declined by 1990 to about 1,200 ha. The most celebrated, and least productive, subvariety is Tschaggele.

The grapes are found in most of the non-varietal light red wines of Trentino-Alto Adige.

🌿 SCHIOPPETTINO, Friuli native in north east Italy. It probably originated in the border area between Prepotto and Slovenia, where wine made from Schioppettino is cited in a marriage ceremony in 1282. In spite of official attempts to encourage its replanting, Schioppettino was substantially neglected after the phylloxera invasion of the late 19th century in favour of the new imports from France. It seemed destined to disappear until an EC decree of 1978 authorized its cultivation in the province of Udine (see also PIGNOLO). The wine is deep in colour, medium in body, but with an attractively aromatic richness hinting at violets combined with a certain peppery quality. Although vine plantings and therefore wine production are still limited, and concentrated in the Colli Orientali, there is clearly potential.

🌿 SCHÖNBURGER, pink-berried 1979 vine crossing with PINOT NOIR, CHASSELAS Rose and MUSCAT HAMBURG among its antecedents which has been more useful to the English wine industry than to its native Germany. There were fewer than 60 ha/150 acres of

SCHWARZRIESLING

Germany planted with this extremely reliable ripener in 1990, most of them in the ever-experimental Rheinhessen and Pfalz, but England's plantings of more than 80 ha made it the country's fifth most planted variety. It has good disease resistance, yields reasonably well, and its tendency to lack acidity is a positive advantage as far from the equator as Kent and Somerset. Its wines are white and relatively full bodied.

SCHWARZRIESLING, or 'black Riesling', is a German synonym for Pinot MEUNIER.

🍇 **SCIACARELLO** (sometimes written **Sciaccarello** and pronounced 'Shackarello'), Corsican speciality capable of producing deep-flavoured if not necessarily deep-coloured reds and fine rosés that can smell of the island's herby scrubland. The vine has good disease resistance and thrives particularly successfully on the granitic soils in the south west around Ajaccio and Sartène. It buds and ripens late and may well have been imported by the Romans but no one has yet identified its Italian cousin. Between 1979 and 1988, the island's total plantings of this characterful variety fell from more than 700 to fewer than 400 ha / 1,000 acres and it is now much less important than NIELLUCCIO.

🍇 **SCUPPERNONG**, oldest and best-known variety of the Muscadine vine species grown for the table and bottle in the southern states of the USA. Like other Muscadines, the grapes (in this case bronze-coloured) and juice are very distinctively flavoured. Some well-structured sweet, dark golden wines are made.

🍇 **SÉGALIN**, recent crossing of JURANÇON Noir × PORTUGAIS Bleu which has good colour, structure, and flavour and is authorized in South West France. (See also CALADOC, CHASAN, PORTAN.)

SEIBEL, common name for many of the French hybrids, many of them having also been given a more colloquially appealing name. Seibel 5455 is more often called PLANTET, for example, while Seibel 4986 is Rayon d'Or and Seibel 9549 is De Chaunac. The variety known as CHANCELLOR in New York State but simply as 'Seibel' in France is Seibel 7053. In the late 1960s there were more than 70,000 ha / 173,000 acres of it planted in France, but it has now almost disappeared from French vineyards. Small quantities of various Seibels are planted in some cooler wine regions around the world.

🍇 **SÉMILLON**, often written plain **Semillon** in non-francophone countries, golden grape from South West France and one of the unsung heroes of white wine production. Blended with its traditional partner SAUVIGNON BLANC, this golden-berried vine variety is the key ingredient in the great dry and sweet whites of Bordeaux, including Sauternes, arguably the world's longest-living unfortified wine, and the best of Graves or Pessac-Léognan.

Unblended, in Australia's Hunter Valley, it is responsible for one of the most idiosyncratic and historic wine types exclusive to the New World. Thanks to its widespread establishment in Bordeaux and much of the southern hemisphere, it has been the world's most planted white grape variety capable of top quality wine production but is not fashionable and is declining in importance.

Sémillon seems destined to play a supplementary role. The wines it produces tend to fatness and, although capable of ageing, have little aroma in youth. Sauvignon Blanc, with its internationally recognized name, strong aroma, high acidity, but slight lack of substance, fills in all obvious gaps. But if Sémillon had traditionally been blended with Sauvignon, it has recently attracted another blend-mate, if for entirely different reasons. Sémillon does not exactly complement Chardonnay so much as provide neutral padding for it and, in a world desperate for Chardonnay, Sémillon has found itself the passive ingredient in commercially motivated blends sometimes, even, called SemChard—most notably but not exclusively in Australia. And here, as elsewhere in the New World, Sémillon's weight, and high yield, make it a popular base for commercial blends.

As a vine Sémillon is easy to cultivate. It is almost as vigorous as Sauvignon Blanc, with particularly deep green leaves, but flowers slightly later and is not particularly susceptible to coulure. It is usefully prey to noble rot but is otherwise disease resistant. This makes Sémillon a particularly productive vine, which was doubtless a factor in its widespread popularity with growers.

Its greatest concentration is still in Bordeaux, where, although total plantings halved between 1968 (when it was the most planted variety of either hue) and 1988, it was still the most planted white grape variety by far with nearly 12,000 ha/30,000 acres in 1992. On the left bank of the Garonne, in the Graves, Sauternes, and its enclave Barsac, Sémillon still outnumbers Sauvignon in almost exactly the traditional proportions of four to one, while in the Entre-Deux-Mers where most Sémillon is planted, Sauvignon (together with varieties for financially more rewarding red wine production) is fast replacing it.

In the great, long-lived dry whites of Graves and Pessac-Léognan Sémillon usually predominates and inspires rich, golden, honeyed, viscous wines quite unlike any Sémillons made elsewhere. Low yields, old vines, oak ageing, and Sauvignon, all play their part. In Sauternes, Sémillon's great attribute is its proneness to noble rot. This special mould, *botrytis cinerea*, concentrates sugars and acids and shrinks yields so that the best of the resulting wines may continue to evolve for centuries. Again, oak ageing deepens Sémillon's already relatively deep gold (really ripe grapes may almost look pink). Thus one of Sémillon's disadvantages, a tendency to over-crop, is eliminated. Similar, but usually less exciting sweet whites, the most ordinary made simply by stopping fermentation or adding sweet grape must, are made in the nearby appellations of Cadillac, Cérons, Loupiac, and Ste-Croix-du-Mont.

In quantitative terms, however, Sémillon's most common expression, other than as basic white in Chile, is as the major ingredient in basic white bordeaux. The best are usually made exclusively of Sémillon with some of the other two 'noble' varieties Sauvignon Blanc and Muscadelle, but up to 30 per cent of the blend may technically comprise Ugni Blanc (the undistinguished TREBBIANO), Colombard, and even the much less common Merlot Blanc, Ondenc, and Mauzac. Cynically made Bordeaux Sémillon can be very dull stuff indeed, high in yield and sulphur content but low in interest, acidity, and flavour.

Like Sauvignon Blanc, Sémillon is allowed in many other appellations for dry and sweet whites of South West France but is perhaps most notable in qualitative terms in Monbazillac. Thanks to its (declining) importance throughout Bergerac, Sémillon is still the most planted variety in the Dordogne, outnumbering Merlot almost two to one. It is also technically allowed in most appellations of Provence, but has made little impact on the vineyards of the Midi where acidity is at a premium.

Sémillon's other great sphere of influence is South America in general and Chile in particular where 2,700 ha/6,700 acres provide rarely exported basic dry white. Argentina has more than 2,000 ha/5,000 acres of Sémillon, which makes it relatively unimportant there.

In North America Sémillon is generally rather scorned, lacking the image of Sauvignon Blanc, although a significant number of producers use the former to add interest to the latter. There are just 2,000 acres/800 ha of usually high-yielding Sémillon in California (where some have experimented with producing botrytized sweet wines in the image of Sauternes from it). It has been given a fillip by the useful part it can play in adding weight to Sauvignon in white bordeaux-like blends called Meritage. Historically Livermore Valley has produced the best Semillon fruit, although it can also perform well in parts of Napa, Sonoma, and the Santa Ynez Valley. Semillon also has a relatively significant presence in Washington, where it often displays grassy, Sauvignon-like aromas.

It is quite widespread, without being particularly important, throughout eastern Europe (especially Croatia), but it is in South Africa and Australia where Semillon has had a particularly glorious past. In 1822, 93 per cent of all South African vineyard was planted with this variety, imported from Bordeaux. So common was it then in fact that it was simply called Wyndruif, or 'wine grape'. It was subsequently called Green Grape, a reference to its abnormally green foliage, but has been declining in importance so that today the **Semillion**, as it is sometimes called, accounts for less than 1 per cent of Cape vineyards.

Semillon is still relatively widely grown in Australia, on the other hand, although it was overtaken by Chardonnay and even Sauvignon Blanc in the late 1980s. It seems to have settled in Australia's wine industry relatively early, possibly having been imported from South Africa, and is still mainly grown in New South Wales, making either extraordinary, ageworthy, full bodied dry whites in the Hunter Valley or more commercial liquids, together with the odd sweet marvel, in the irrigated vineyards inland. Only the best bottles from the Hunter and Bordeaux demonstrate Semillon's ability to age and, often, its tendency to acquire an almost orange depth of colour when it does. It was not until the 1980s, however, that Semillon was publicly revealed as the source of the Hunter's greatest. Until then the wines were usually called 'Hunter Riesling', and occasionally 'Chablis' and 'White Burgundy' depending on slight variations in style.

In Australia's cooler sites such as Tasmania and the south of Western Australia, Semillon often demonstrates the same sort of grassiness as in Washington State, and in New Zealand there are a few hundred hectares. Some interesting sweet wines have been coaxed out of Semillon in Gisborne.

The variety was also exported to what is now Israel to establish vineyards there at the end of the 19th century.

SERCIAL, Anglicized name for the Portuguese white grape variety CERCEAL, once quite commonly planted on the island of Madeira but subsequently used to denote the lightest, driest, most delicate style of madeira rather than the grape variety from which it was originally made. The vine ripens particularly late and grapes retain their acidity quite notably (that rarity, pure varietal Sercial madeira, can take many decades to reach maturity). In the post-phylloxera era it has been easier to find the variety on the mainland, as ESGANA CÃO, than on the island, although plantings are now increasing. Rating as a madeira given.

SERVANIN, old Isère vine heading for extinction.

SEVERNY, Russian vine variety developed at the all-Russia Potapenko Institute from a Malengra seedling with a member of the famously cold-hardy *Vitis amurensis* vine species native to Mongolia. See also SAPERAVI Severny and CABERNET SEVERNY. There is small-scale experimentation with a Severny variety in Canada, where winters are as harsh as in Russia.

SEYVAL BLANC, SEYVE VILLARD 5276, most important of the many French hybrids created by Bertille Seyve and his partner and father-in-law Victor Villard in France in the early 20th century. It is the result of crossing two SEIBEL hybrids and, for growers in cool regions such as England and New York State, is usefully early budding and ripening. It is the most planted variety in England after MÜLLER-THURGAU and grows in every NY wine region. It can be aged in oak with malolactic fermentation, or it can be made in stainless steel for a clean, fruity style. The grape is fairly neutral but has no trace of any non-*vinifera* flavour.

SEYVE-VILLARD, series of French hybrids developed by hybridizer Villard from the hybrid Joannes Seyve. France had a total of more than 3,000 ha/7,500 acres of various Seyve-Villard hybrids, most of them making red vin de table, planted in the late 1980s. SEYVAL BLANC is the most famous.

🍷 **SHIRAZ,** Australian (and South African) name for the SYRAH grape, and as such a name arguably better known by consumers than its Rhône original. (Italian varietal Syrah has been known to be labelled Shiraz for purely commercial reasons.) French wine producers are typically reticent about identifying grape varieties, however noble. Shiraz, then known as Scyras, was probably taken to Australia, possibly from Montpellier, in 1832 by James Busby. It flourished so obviously that it was rapidly adopted by New South Wales and spread like wildfire therefrom.

The ubiquity of Shiraz in Australia (its 7,000 ha/17,500 acres in 1994 represented almost a third of the nation's total black grape vineyard) did little to imbue Syrah with the respect it commands in its homeland and during the 1970s Cabernet Sauvignon was much more highly regarded. A new-found appreciation of the quality of wine produced from older Shiraz vines in unirrigated areas, especially the Barossa Valley, has since propelled Shiraz back into the limelight. It has had particular success in the Barossa Valley in South Australia, the Hunter Valley in New South Wales, and in a number of wine areas in the state of Victoria where the cooler conditions can result in more peppery styles. Today Australian Shiraz can vary from a brown, baked, dilute everyday red to the glorious, almost porty concentration of the likes of Australia's most famous wine, Penfolds Grange. Like top quality Hermitage, this Syrah-dominated wine can taste oddly like Bordeaux after 20 years in bottle. Viticulturally Shiraz is identical with Syrah but the resulting wines are very different, with Australian versions tasting much sweeter and riper, more suggestive of chocolate than the pepper and spices often associated with Syrah in the Rhône.

Shiraz appears on possibly the majority of Australian red wine labels, either in lone varietal splendour or in conjunction with, most often, Cabernet Sauvignon—typically labelled simply Cabernet Shiraz or Shiraz Cabernet, depending on which is the dominant variety.

SHIROKA MELNISHKA, see MELNIK.

🍷 **SIDERITIS,** Greek variety found to a limited extent near Patra. Often blended with RODITIS.

🍷 **SIEGERREBE,** modern German crossing grown, like certain giant vegetables, by exhibitionists. In Germany it can break, indeed has broken, records for its ripeness levels, but the flabby white wine it produces is so rich and oppressively flavoured that it is a chore to drink. It was bred from GEWÜRZTRAMINER and a red table grape and has been known to reach *double* the ripeness reading required for a Trockenbeerenauslese. It is, however, one of the very few modern German crossings that is not particularly productive ('only' 40 to 50 hl/ha (2.3–2.8 tons/acre) as it is particularly susceptible to coulure) and is useful only as an enriching, very minor ingredient in a blend. So powerful is its heady flavour that a fine Riesling is quite overpowered by even 10 per cent of Siegerrebe. Total German plantings were still more than 200 ha/500 acres in 1990, mainly in Rheinhessen and Pfalz, but are expected to decline. The variety can usefully bolster some blends in England.

🍇 SILVANER, once important German grape grown widely in central Europe. Its very name suggests romantic woodland origins. It has a long history over much of eastern Europe, where it may indeed first have been identified growing wild. Its origins are thought to be in what is now Austria (although it is hardly grown there today) and it came to Germany from the banks of the Danube (Österreicher was once a common synonym for it) although some posit Transylvanian origins. It probably invaded Germany from the east, and a vine known as Silvaner was widely grown throughout the extensive vineyards of medieval Germany. Its arrival from Austria in the Franken region in 1659 is well documented and the variety is still the second most planted in this German region. Silvaner enjoyed its greatest popularity in the first half of the 20th century when it took over from ELBLING as Germany's most planted vine variety, before that position was taken over by the more productive, less geographically fastidious, and even earlier ripening MÜLLER-THURGAU. It can produce much more racy wine than this dreary usurper. In Germany it is often called Grüner Silvaner, whereas in France (as in Austria) it is known as SYLVANER.

This vigorous vine buds a few days before Germany's quintessential RIESLING, and can suffer spring frost damage. It is not notable for its disease resistance but it is productive. The chief characteristic of the wine produced is its high natural acid, generally lower than Riesling's in fact but emphasized by Silvaner's lack of body and frame. (It is significant that the German name for SAUVIGNON BLANC, a variety essentially notable for its aroma and high acid is Muskat-Silvaner.) Provided yields are not too high, it can provide a suitable neutral canvas on which to display more geographically based flavour characteristics, but wine made from Silvaner is noted neither for its longevity nor its natural ripeness.

Most of Germany's finest, earliest, most crackling, Silvaners come from Franken, where Riesling is difficult to ripen and Silvaner has remained popular. Here the variety sometimes called Franken Riesling (in both Germany and California) is capable of tingling concentration, and even some exciting late harvest sweet wines. Elsewhere, Silvaner is being systematically replaced by more fashionable varieties. About half of the country's Silvaner is planted in the Rheinhessen (where there are some particularly suitable sites in the north) with much of the rest in Franken and Pfalz.

Prized for its reliable early ripening and productivity, Silvaner is a parent in a wide variety of Germany's army of modern vine crossings, including BACCHUS, EHRENFELSER, MORIO-MUSKAT, OPTIMA, RIESLANER, and SCHEUREBE. It is well adapted to the German climate but is less versatile in choice of site than Riesling, a more profitable choice in many cases.

Outside Germany Silvaner is relatively important in Slovenia and is grown all over eastern Europe. As **Sylvaner Verde**, it is also planted in Alto Adige, the Italian Tyrol, where it can provide light and piercing wines for youthful consumption.

In the Valais of Switzerland it is the second most planted grape and is called Johannisberg, Gros Rhin, or just Grüner. Its wines can seem positively luscious in comparison to French Switzerland's ubiquitous CHASSELAS, which ripens before it.

Despite its useful acidity levels Silvaner is not widely grown in the New World, although California still grew 185 acres of 'Sylvaner', mainly in Monterey, in the early 1990s and there were nearly 120 ha / 300 acres of 'Sylvaner' grown in Australia.

Blauer Silvaner is a local, dark-berried mutation that is a speciality of Württemberg.

SIRAH, name occasionally misleadingly used for some PETITE SIRAH in South America. It should not be confused with the true SYRAH of the northern Rhône.

🍇 SMEDEREVKA, quantitatively important grape in the south of what was Yugoslavia, where it was the second most important white grape after WELSCHRIESLING before war broke out in 1991. It takes its name from the town of Smederevo south of Belgrade and is planted extensively in Serbia and Vojvodina. As a wine, it is usually dry and relatively high in both alcohol and acidity, but it is often blended with other varieties. It is also planted to a much more limited extent in Hungary.

🍇 SOUSÃO, or SOUZÃO, widely planted in Portugal's Douro valley, where it is regarded as a useful, if slightly rustic, ingredient in port for its colour and obvious fruit character in youth. It has also been planted by aspirant makers of port style wines in California and Australia, with a certain degree of success.

🍇 SOUSÓN, minor Galician grape.

🍇 SPAGNOL, ancient Provençal vine grown around Nice.

SPANNA, local name for the NEBBIOLO grape in eastern Piedmont in north west Italy, particularly around Gattinara.

SPÄTBURGUNDER, chief German synonym for PINOT NOIR.

SPÄTROT, synonym for ZIERFANDLER.

SPERGOLA, occasional name for SAUVIGNON BLANC in Italy's Emilia region.

STEEN, common South African name for a slightly soft varietal version of the CHENIN BLANC that is so widely planted there.

STRACCIA CAMBIALE, synonym for BOMBINO BIANCO.

SUBIRAT PARENT, local name for MALVASIA Riojana in Penedès, where the produce of about 300 ha/750 acres is used to perfume blends.

🍇 **SULTANA,** more common name for THOMPSON SEEDLESS outside California. This versatile grape is widely grown, especially in the Middle East (where it originated) and Australia. Also known as **Sultanine** and **Sultanina,** and variants of Kismis, it is still mainly grown for drying and for table grapes but in Australia and California some of the crop has been diverted to answer the needs of the wine industry in times of shortage. It cannot make interesting wine but, with careful vinification, can certainly make serviceable, if fairly neutral, table wine. It is also widely planted in South Africa, where it is the second most important variety after CHENIN BLANC. Most of its produce is either distilled or sold as grape concentrate here, however.

🍇 **SUMOLL,** minor Spanish grape grown in the Conca de Barberá zone.

🍇 **SUSUMAIELLO,** lively, deep-coloured red grape that has crossed the Adriatic to be grown on the heel of Italy.

SVATOVAVRINECKE, Czech name for ST LAURENT.

🍇 **SYLVANER,** the French name for the grape spelt Silvaner in German. In France it is virtually exclusive to Alsace, where it is still the most planted vine in the lower, flatter, more fertile vineyards of the Bas-Rhin (although Riesling and to a lesser extent Gewürztraminer are catching up fast). The total area planted with Sylvaner, about 2,700 ha/6,700 acres, has been one of the few constants in Alsace in the late 20th century.

Sylvaner may be an old vine and, at one time an extremely important one in Germany at least, but most of the wines it produces in Alsace are decidedly dull. They generally have good body and acid, but tend to lack much specific flavour unless planted on a particularly suitable site (which would probably produce even better RIESLING). It is worth ageing only when made in the leanest of styles, such as that of Trimbach.

See SILVANER for more details.

🍇 **SYMPHONY,** California crossing of GRENACHE Gris and MUSCAT OF ALEXANDRIA developed at Davis by Dr H. P. Olmo. In the early 1990s it enjoyed a small vogue as an off-dry table wine, similarly to MALVASIA Bianca, and as a sparkling wine. Of the state total of more than 200 acres/80 ha, the most successful plantings have been in Sonoma.

Syrah

♥ **SYRAH,** one of the noblest black grapes, if longevity is a factor
in nobility. The great grape of the northern Rhône, Syrah's origins
are the subject of much debate and hypothesis. The vine is
relatively productive and disease resistant, sensitive to coulure,
but conveniently late budding and not too late ripening. The deep,
dark, dense qualities of the wine, characteristically and strangely
satisfying scented with black pepper and burnt rubber, are much
reduced once the yield is allowed to rise and Syrah has a tendency
to lose aroma and acidity rapidly if left too long on the vine.

Many *vignerons* in the northern Rhône distinguish between a
small-berried, superior version of Syrah, which they call Petite
Syrah (although there is no relation between Syrah and the variety
known in North and South America as PETITE SIRAH), and the
larger-berried Grosse Syrah, which produces wines with a lower
concentration of phenolics. Ampelographers reject this distinction,
although connoisseurs have reason to be grateful for it. The total
pigments in Syrah can be up to 40 per cent higher than those in
the tough, dark CARIGNAN, which makes it, typically, a wine for the
long term that responds well to oak maturation, even new oak at
its ripest.

The most famous prototype Syrahs—Hermitage and Côte
Rôtie—are distinguished by their longevity or, in the case of newer
producers, ambition. Of North Rhône Syrahs, only St-Joseph and
Crozes-Hermitage can sensibly be broached within their first five
years. And the better wines of Cornas repay bottle maturation
even more handsomely than some Hermitages. Syrah that has
not reached full maturation can be simply thin, acrid, and
astringent. When planted on the fringes of the Rhône such as in
the Ardèche, Syrah may avoid this fate in only the ripest vintages.

Until the 1970s French Syrah plantings were almost exclusively
in and around the very limited vineyards of the northern Rhône
Valley and were dwarfed in area by total Syrah plantings in the
vine's other major colony, Australia, where it is known as SHIRAZ.
Since then, however, Syrah has enjoyed an extraordinary surge in
popularity throughout southern France so that total French
plantings rose from 2,700 ha/6,700 acres in 1968 to exactly 10
times that 20 years later. The increases were noticeable
throughout the southern Rhône, particularly in Châteauneuf-du-
Pape country, where Syrah was being increasingly valued as
endowing Grenache with life expectancy, but have been most
spectacular to the west in the Languedoc, especially in the Gard
and the Hérault, where Syrah has been most enthusiastically
adopted as an officially approved 'improving variety' that has
added structure to both Coteaux du Languedoc wines and vins de
pays. By 1993 there were nearly 24,000 ha/60,000 acres of Syrah
in the Languedoc-Roussillon, making it the fifth most planted vine
after, in order, CARIGNAN, GRENACHE, CINSAUT, and ARAMON. Syrah
has frequently been responsible for the Midi's most successful
varietals, usually labelled Vin de Pays d'Oc. Yields very much in
excess of the low yields that characterize the arid hill of Hermitage

have somewhat diluted its North Rhône characteristics, however. In the North Rhône it is rarely blended, except perhaps for a little VIOGNIER, while in the south it is typically blended with Grenache and perhaps Mourvèdre and Cinsaut. In Provence the very Australian blend of Syrah and Cabernet Sauvignon is becoming more common and Syrah is one of the most successful noble vine imports to Corsica, where there are more than 200 ha / 500 acres in production.

Another unexpectedly successful site for mature, concentrated Syrah is the Valais in Switzerland, particularly around the suntrap village of Chamoson in the upper reaches of the Rhône Valley. Here classic North Rhône techniques are employed, sometimes to great effect. Spain's first commercial Syrah was released in 1995, from vineyards near Toledo. Italy too is flirting with Syrah, most successfully so far at Isole e Olena in Tuscany.

To ripen fully Syrah demands a warm climate, which naturally limits its spread, but some plantings in California have been very successful. Californians were slow to distinguish between true Syrah and Petite Sirah and even slower to import suitable plant material so that, despite the modishness attached to all things Rhôneish, there were still barely 400 acres / 160 ha of it in the state in 1992. Washington State clearly has Syrah potential.

South Africa's total had reached nearly 900 ha / 2,200 acres by 1992, most of it in Paarl and Stellenbosch, and the results are promising in those who manage to restrict yields. A small amount is also grown with some success in Chile and in Argentina, where it has also been known as Balsamina, not a bad synonym for this headily scented grape variety.

SZÜRKEBARÁT, Hungarian name for PINOT GRIS quite widely planted there, but its naturally low acidity makes for slightly flabby wines, particularly on the Great Plain. It is most revered within Hungary as Badacsonyi Szürkebarát, a rich, heavy wine from the north shore of Lake Balaton. It can yield livelier wines from the Mátra Foothills.

T

TALIA. See THALIA.

TAMAREZ, common Portuguese name for a variety of white grapes, around which is still some confusion.

TĂMÎIOASĂ, Romanian name for MUSCAT. Thus **Tămîioasă Alba** is Romanian for MUSCAT BLANC À PETITS GRAINS, **Tămîioasă Hamburg** or **Tămîioasă Neagră** is MUSCAT HAMBURG, **Tămîioasă Ottonel** is MUSCAT OTTONEL. See also below.

Tămîioasă Românească

🍇 **Tămîioasă Românească,** old Romanian variety, of which there were more than 1,000 ha/2,500 acres planted in Romania in the early 1990s. It is a very powerfully, aristocratically scented grape variety well suited to producing sweet white wines of real distinction. Also grown in Bulgaria as **Tamianka**. In Germany it is known tellingly as the Weihrauchtraube, or 'frankincense grape'.

🍇 **Taminga,** relatively new variety bred specifically for Australian conditions (see also TARRANGO) by A. J. Antcliff. Taminga is capable of ripening and producing large quantities of commercially acceptable wine in a wide variety of different sites.

Tamyanka, Russian synonym for the refined MUSCAT BLANC À PETITS GRAINS.

🍇 **Tannat,** tough variety most famous as principal ingredient in Madiran, where its inherent astringence is mitigated by blending with CABERNET FRANC, some CABERNET SAUVIGNON and FER, and wood ageing for at least 20 months. Young Tannat can be so deep coloured and tannic that it recalls NEBBIOLO. The wine is spicy, mouth-filling, and exciting. The variety is also blended into Côtes de St Mont, Irouléguy, Béarn, and the rare reds and pinks of Tursan.

Although it can also be found as a minor ingredient in such wines as Côtes du Brulhois, overall plantings in France have been declining so that there were fewer than 3,000 ha/7,500 acres by 1988. Although it may owe its French name to its high tannin content, the vine is almost certainly Basque in origin and, like MANSENG, was taken to Uruguay by Basque settlers in the 19th century. There are still several thousand ha there, where it is called Harriague, presumably after its original promulgator. From here it spread to Argentina, where it is still grown to a very limited extent.

🍇 **Tarrango,** 1960s Australian crossing of TOURIGA × SULTANA to provide a slow-ripening variety suitable for the production of light, soft, fast-maturing, relatively crisp reds somewhat in the style of Beaujolais. Tarrango will ripen satisfactorily only in the hot irrigated inland wine regions of Australia. Brown Brothers of Milawa have been particularly persistent with it. Other varieties developed by A. J. Antcliff specifically for Australian conditions included Carina and Merbein Seedless for drying and Tulillah, Goyura, and TAMINGA for white wines.

🍇 **Tazzelenghe,** tongue-stingingly tart speciality of the far north east of Italy, notably the Colli Orientali.

Teinturier, French term for any red-fleshed grape. **Tintorera** is the Spanish equivalent.

🍇 **Tempranillo,** Spain's answer to CABERNET SAUVIGNON, the grape that puts the spine into so many good Spanish reds. Its grapes are thick-skinned and capable of making deep-coloured, long-lasting wines that are not, unusually for Spain, notably high in alcohol.

Temprano means early in Spanish and Tempranillo probably earns its name from its propensity to ripen early, certainly up to two weeks before the Garnacha (GRENACHE) with which it is almost invariably blended to make its most important wine, Rioja. This relatively short growing cycle (Tempranillo buds neither early nor late) enables it to thrive in the often harsh climate of Rioja's higher, more Atlantic-influenced zones Rioja Alta and Rioja Alavesa, where it constitutes up to 70 per cent of all vines planted. Tempranillo has traditionally been grown in widely spaced bushes here, but this relatively vigorous, upright vine has responded well to recent efforts to train it with more rigour on wires.

Wine made from Tempranillo grown in relatively cool conditions, where its tendency to produce musts slightly low in acidity is a positive advantage, can last well. Grape flavours can vary from tobacco leaves to spice and leather, but many Tempranillo-based wines are flavoured at least as much by oak as by fruit. A high proportion of Tempranillo is blended with juicier grapes: in Rioja with Garnacha, Mazuelo (CARIGNAN), and Viura (MACCABÉO). In Penedès, where it is known as Ull de Llebre and Ojo de Liebre in Catalan and Spanish respectively, Tempranillo stiffens and darkens the local Monastrell (MOURVÈDRE). In Valdepeñas, where, as Cencibel, Tempranillo is the dominant black grape, white grapes are commonly added to soften the wine, which is not always desperately noble or long-lasting here, and is often flavoured with oak chips. It is also grown in La Mancha, Costers del Segre, Utiel-Requena, and increasingly in Navarre and Somontano. The variety's perfect spot, however, seems to be the cool Ribera del Duero region north of Madrid, where, as Tinto Fino, it is by far the principal grape variety. Here it produces wines loaded with colour, tannins, acidity, and flavour, but even so it may be seasoned with CABERNET SAUVIGNON, GRENACHE, ALBILLO.

The variety is so well entrenched in northern and central Spain that Spain's total Tempranillo plantings were 87,800 ha/217,000 acres in the early 1990s, making it the country's fourth most important black grape after Garnacha, Monastrell, and BOBAL.

Such has been the success of Ribera del Duero's Vega Sicilia and Pesquera wines that producers outside Spain have been taking note of Tempranillo. There were already 2,500 ha of the variety by the late 1980s over the Pyrenees in the Languedoc-Roussillon, where Tempranillo is officially recommended (though not for appellation contrôlée wines), notably in the Aude.

Tempranillo is one of the very few Spanish varieties to have been adopted in any quantity (about 5,200 ha/13,000 acres) in Portugal, where it is known as (Tinta) Roriz and is a valued, if not particularly emphatic, ingredient in port blends. Also in the Douro, downstream of Ribera del Duero, it has demonstrated a real aptitude for table wines. Tinta Roriz, also known as Tinta Aragonez, is also grown increasingly in Dão.

As **Tempranilla** and making rather light, possibly over-irrigated reds, it has been important in Argentina's wine industry but is losing ground to more marketable grape varieties and in 1990 covered 7,500 ha, mainly in Mendoza.

Tempranillo may well be the true identity of the unfashionable low-acid variety known in California as Valdepeñas, of which 500 acres/200 ha remained, mainly in the Central Valley, in 1992.

Tempranillo's starring role in one of the world's most famous and distinctive wines, Rioja, makes it a likely ingredient in any of the world's many experimental vine nurseries. This is a variety which is expected to expand its sphere of commercial influence.

♀ **TÉOULIER,** old south east French vine which may have originated in Manosque, to judge from its synonyms. Rarely found today.

TERAN, name for a subvariety of the red Friuli grape variety REFOSCO.

TERMENO AROMATICO. See GEWÜRZTRAMINER.

♂ **TEROLDEGO ROTALIANO,** sometimes just **Teroldego,** north Italian grape capable of producing deep-coloured, seriously lively, fruity wines with relatively low tannins for early drinking. It performs this trick almost exclusively on the Rotaliano plain in Trentino.

TERRANO, alternative name for Teran, or REFOSCO.

♀ **TERRANTEZ,** now, unfortunately, almost extinct but once a treasured Madeira variety making rich, perfumed wines.

TERRET, one of the Languedoc's oldest vines which, like PINOT, has had plenty of time to mutate into different shades of grape which may even be found on the same plant. Indeed Galet claims to have seen different-coloured grapes in the same bunch. All Terrets bud usefully late and keep their acidity well.

♀ **TERRET GRIS,** widely planted on the plains of the Languedoc, where there were well over 5,000 ha/12,300 acres planted in 1988. It can be made into a relatively full bodied, sometimes mineral-scented, naturally crisp varietal white, but as a name Terret lacks the magic of internationally known varieties. Even as recently as 1993, the light-skinned Terrets were at least as common in the Languedoc as CHARDONNAY. Although light-berried Terrets (rarely distinguished by officialdom) are in decline, combined they represented France's ninth most planted white wine variety at the end of the 1980s. They are allowed into the white wines of Minervois, Corbières, and, to a decreasing extent, Coteaux du Languedoc.

Terret Blanc was once grown near Sète expressly for the local brandy and vermouth industry (it too was prized for its acidity).

❦ **TERRET NOIR**, is the dark-berried version of TERRET GRIS which is grown on a much more limited scale (hardly 800 ha/2,000 acres remained in 1988) but is one of the permitted varieties in red Châteauneuf-du-Pape to which it can add useful structure and interest.

THALIA, Portuguese name for the ubiquitous UGNI BLANC.

❦ **THINIATIKO**, occasionally found on the Greek island of Cephalonia making rich wines. May be related to MAVRODAPHNE.

THOMPSON SEEDLESS, common California name for the seedless white SULTANA grape used more for dried fruit than for wine. William Thompson was an early grower of the variety, near Yuba City. It is California's most planted grape variety by far, its total of 270,000 acres/110,000 ha in 1994 dwarfing even Chardonnay's total California plantings of 69,600 acres. Almost all of California's Thompson Seedless is grown in the hot, dry Central Valley, with nearly two-thirds in Fresno County, the powerhouse of California raisin production, alone. In the 1970s Thompson Seedless was particularly useful to the California wine industry in helping to bulk out inexpensive jug wine blends at a time when demand far outstripped supply of premium white wine grape varieties.

❦ **TIBOUREN**, *the* Provençal grape variety with a long history and the ability to produce such quintessentially Provençal wines as earthy rosés with a genuine scent of the *garrigue*. Planted in strictly limited quantities, it is sensitive to coulure and therefore yields irregularly. The deeply incised shape of its leaves reminds Galet of some Middle Eastern vine varieties, and certainly it could possibly have been imported by the Greeks via Marseilles.

TINTA, TINTO, means literally 'dyed' in Spanish and Portuguese and is widely used as a prefix for dark-skinned grape names.

❦ **TIMORASSO**, Piedmontese rarity making aromatic, durable wine, and grappa.

❦ **TINTA AMARELA**, productive Portuguese vine once grown more extensively in the Douro and still also grown in Portugal's Dão region. It can yield attractively scented wines but is not regarded as one of the finest port varieties and is particularly sensitive to rot.

TINTA BAIRRADA, occasional name for BAGA.

❦ **TINTA BARROCA**, sturdy port grape which does well in the harsh climate of Portugal's Douro Valley. Yields are relatively good and the vine can be depended upon to produce useful, dark-skinned grapes even in drought conditions. Perhaps because of this, Tinta Barroca (often spelt **Tinta Barocca**) has been the most popular port variety in South Africa's vineyards and full-throttle, unfortified varietal Tinta Barroca dry(ish) red is a South African speciality. In the Douro the wine is treasured for its delicacy and sweet, gentle fruity aroma.

🍇 **TINTA CARVALHA**, widely planted (because productive) Portuguese grape producing very thin, ordinary reds.

TINTA DE TORO, another name for TEMPRANILLO.

🍇 **TINTA FRANCISCA**, another port grape. The wine produced can be notably sweet but is not particularly concentrated.

🍇 **TINTA MADEIRA**, vine planted in some of California's hottest vineyards and used for port-style wines.

🍇 **TINTA NEGRA MOLE**, by far the most commonly planted vine variety on the island of Madeira. Although its background is unknown, Negra Mole is a *vinifera* variety (unlike many of the vines that replaced the noble varieties SERCIAL, VERDELHO, BUAL, and MALVASIA after the ravages of downy mildew and phylloxera in the 19th century). It yields relatively high quantities of very sweet, red wine which turns amber quite rapidly thanks to the madeira production process and then yellow/green with age. A vine of the same name, which may be quite distinct, is also grown on the Portuguese mainland in the Algarve and, as Negramoll, in Spain.

🍇 **TINTA PINHEIRA**, ordinary Portuguese grape variety used for making red Dão.

TINTA RORIZ, see RORIZ.

TINTO DEL PAIS, TEMPRANILLO synonym.

🍇 **TINTO CÃO**, top quality port grape which almost disappeared from the Douro valley in northern Portugal (despite its long history there) but is now cherished as one of the five finest port varieties, albeit producing spicy wine which is not especially deep-coloured and is apt to oxidize. It gives best results when planted in cooler spots. It has also been planted experimentally at Davis in California.

TINTO FINO, local name for the well-adapted TEMPRANILLO vine in Spain's Ribera del Duero region.

🍇 **TOCAI**, or **TOCAI FRIULANO**, the most popular and widely planted white Friuli grape in north east Italy. It has no connection at all with TOKAY D'ALSACE (an Alsace synonym for the PINOT GRIGIO which grows alongside Tocai in Friuli), and is probably completely unrelated to the famous Hungarian dessert wine Tokaji. In Italy in 1990 there were a total of nearly 7,000 ha/17,300 acres planted with Tocai Friulano.

In Friuli this productive, late-budding vine variety produces the staple wine of the region's taverns and *trattorie*, with more than

2,000 ha currently in production in the major Colli Orientali, Collio, Grave del Friuli, and Isonzo DOC zones. The wine is light in colour and body, and can be anything from floral and crisp to almondy in flavour. It is designed to be drunk young.

Some Tocai, not identical to Tocai Friulano, is also grown in the Veneto.

Some Italian authorities, notably Gaetano Perusini, assert that the variety is closely related to the FURMINT grape of Tokay and was imported from Hungary by Count Ottelio di Ariis in 1863 at a time when Friuli was part of the Austro-Hungarian empire, while Friuli's historian Coronini claims that cuttings of Tocai were sent to King Bela IV of Hungary by Bertoldo di Andechs, patriarch of Aquileia, in the first half of the 13th century. The dispute remains undecided but, under pressure from Hungary, the Italians have agreed eventually to give up using the name Tocai.

According to Galet, however, it is identical to SAUVIGNON VERT or Sauvignonasse, the variety planted in much of Chile. A variety called Tocai Friulano is also cultivated to a limited extent in Argentina, while a Sauvignon Vert is grown in Ukraine.

TOKAY D'ALSACE, or simply **TOKAY**, was for long the Alsace name for PINOT GRIS but has been outlawed under pressure from Hungary. Tokay is also used elsewhere in France, and Australia, as a synonym for MUSCADELLE.

TORBATO, Sardinian speciality not widely planted. Like GRENACHE Noir, known as Cannonau in Sardinia, its origins are disputed, thanks to the past extent of the kingdoms of Majorca and Aragon. Many believe it to be a Spanish variety which was imported many centuries ago. It is particularly successful around Alghero.

It was once quite widely cultivated in Roussillon, southern France, where it is known as Tourbat, or Malvoisie du Roussillon but was abandoned before new and healthier plant material was imported from Sardinia in the 1980s. The wine can be scented and quite full bodied.

TORONTEL, **TORONTÉS**, and **TORRONTÉS**, white grape variety or varieties gaining increasing recognition in the Spanish-speaking world.

Torrontés is the name of a distinctively flavoured indigenous variety in Galicia in north west Spain that is particularly common in the white wines of Ribeiro. Within Spain the variety is occasionally found around Cordoba. Much more important, however, are several white grape varieties known as Torrontés in Argentina, where they were planted on a total of nearly 16,000 ha/39,500 acres in 1990. Although there was considerable emigration from Galicia to Argentina, no definite relationship between Spanish and Argentine Torrontés has been established. Some regard Torrontés as the Argentinian white wine variety with the greatest potential. Carefully vinified, Torrontés can produce wines for early drinking that are not too heavy, are high in acidity, and are intriguingly aromatic in a way reminiscent of but not identical to MUSCAT, although much is also used for blending. The variety seems particularly well adapted to the arid growing conditions of Argentina, where its high natural acidity and assertive flavour distinguish it respectively from the PEDRO GIMÉNEZ and UGNI BLANC which still cover a greater area of vineyard.

Torrontés Riojano is the most common Argentine subvariety and takes its name from the northern province of La Rioja, where it is by far the most planted single vine variety. **Torrontés Sanjuanino** is more commonly associated with the province of San Juan in Argentina and is rather less widely planted. It is less aromatic, and has bigger berries and more compact clusters. In Chile it is known as MOSCATEL DE AUSTRIA. Argentine vineyard statistics also distinguish the relatively rare **Torrontés Mendocino**, sometimes called **Toróntes Mendozino**, which is actually most common in Río Negro province in the south and lacks Muscat aroma.

There are also several hundred ha of a variety known as **Torontel** in Chile.

TOURBAT, Roussillon name for TORBATO.

TOURIGA is used as an Australian synonym for TOURIGA NACIONAL, but the Touriga of California is probably TOURIGA FRANCESA.

🍇 **TOURIGA FRANCESA**, robust and fine port grape widely grown in Portugal's Douro Valley and the Trás-os-Montes wine region. Despite its name, it is thought to be Portuguese in origin rather than French. It is classified as one of the best port varieties, although the wine it produces is not as concentrated as that of the rarer TOURIGA NACIONAL and is not so highly regarded as that of TINTA BARROCA and RORIZ. Its wines are notable for their perfume and persistent fruit. It should not be confused with the much less distinguished Portuguese variety TINTA FRANCISCA.

🍇 **TOURIGA NACIONAL**, the most revered grape for port, producing small quantities of very small berries in the Douro Valley and, increasingly, the Portuguese Dão region which result in deep-coloured, intensely flavoured, very tannic, concentrated wines. The vine is vigorous and robust but may produce just 300 g / 10 oz of fruit per vine. Newer clones are slightly more productive and average sugar levels even higher. Touriga Nacional should consititute at least 20 per cent of all red Dão, although the wine's suitability as a vintage port ingredient are more obvious than as a red table wine. The variety is also planted to a limited extent in Australia, where it is known simply as **Touriga** and is used to add finesse to fortified wines.

🍇 **TRAJADURA,** delicate, scented variety used to add body and alcohol to Portugal's Vinho Verde and known as TREIXADURA across the Spanish border in Galicia. It is often blended with LOUREIRO and occasionally with ALVARINHO.

🍇 **TRAMINER, TRAMINI, TRAMINAC,** the less aromatic, paler-skinned progenitor of the pink-skinned white wine grape variety GEWÜRZTRAMINER. It has been grown, for example, in Moravia in what was Czechoslovakia, where it is also known as Prinç. The name derives from the town of Tramin, or Termeno, in the Alto Adige. In countries as different as Germany, Italy, Austria, Romania, much of the former Soviet Union, and Australia, however, Traminer is also used as a synonym for Gewürztraminer, under which name more details can be found. See also SAVAGNIN.

🍇 **TRBLJAN,** relatively important variety within former Yugoslavia, particularly on the Croatian coast just north of Zadar. A total of more than 3,000 ha/7,400 acres are planted. The grape is also sometimes called Kuč.

🍇 **TREBBIANO,** most common name for the undistinguished UGNI BLANC in Italy, where it is by far the most planted white grape variety. The word Trebbiano in a wine name almost invariably signals something light, white, crisp, and uninspiring. This gold-, even amber-berried grape variety is so prolific, and so much planted in both France and Italy, the world's two major wine-producing countries, that it probably produces more wine than any other vine variety in the world—even though Spain's AIRÉN and dark-skinned Garnacha/GRENACHE cover a larger total vineyard area. In Italy Trebbiano in its many forms covers a greater area of vineyard even than SANGIOVESE. It is cited in more DOC regulations than any other single variety (about 80) and may well account for more than a third of Italy's entire DOC white wine production.

In France, where it found its way as a result of the Mediterranean trade that flourished between Italian and French ports during the 14th century, the variety is the country's most important white vine. See under its most common French name UGNI BLANC for details of Trebbiano in France.

Ugni Blanc's most common use, as base wine for brandy, provides a clue to the character of the wine produced by Trebbiano. It is, like most copiously produced wines, low in extract and character, relatively low in alcohol, but usefully high in acidity. This exceptionally vigorous vine buds late, thereby avoiding most spring frost damage. It gives naturally high yields, which can easily reach 150 hl/ha (8.5 tons/acre). It has good resistance to powdery mildew and rot but can succumb to downy mildew. Because it ripens relatively late, often as late as October in parts of Italy, there is a natural geographical limit on its cultivation, but in areas such as Charentes it is simply picked before being fully ripe, as indeed it is in southern Italy to maximize acidity.

TREBBIANO

There are almost as many possible histories of Trebbiano as there are identified subvarieties in Italy. **Trebbiano Toscano** (Tuscan) and **Trebbiano Giallo** (yellow) ripen rather earlier than most other Trebbiano subvarieties. Today Trebbiano is planted all over Italy (with the exception of the cool far north), to the extent that it is likely that the great majority of basic Italian whites contain at least some of the variety, if only to add acidity and volume. Its stronghold, however, is central Italy. Trebbiano Toscano, covering almost 60,000 ha / 148,000 acres, was Italy's third most planted vine variety in 1990, while there were more than 20,000 ha of **Trebbiano Romagnolo**, nearly 12,000 ha of **Trebbiano d'Abruzzo**, nearly 5,000 ha of Trebbiano Giallo, and more than 2,000 ha of **Trebbiano di Soave**.

Trebbiano di Romagna dominates white wine production of Emilia-Romagna, made chiefly from Trebbiano Romagnolo or the almost amber-berried **Trebbiano della Fiamma**. Some idea of its ubiquity is given by listing just some of the wines in which it is an ingredient: Verdicchio, Orvieto, Frascati, together with Soave from Trebbiano di Soave and Lugana from **Trebbiano di Lugana**. The variety has after all had many centuries to adapt itself to local conditions. Between Tuscany and Rome, in Umbria, the variety is known as Procanico, which some agronomists believe is a superior, smaller-berried subvariety of Trebbiano. Only the fiercely varietal-conscious north eastern corner of Italy is virtually free of this bland ballast.

Trebbiano's malign influence was most noticeable in Central Tuscany in much of the 20th century, when Trebbiano was so well entrenched that Chianti and therefore Vino Nobile di Montepulciano laws sanctioned its inclusion in this red wine, thereby diluting its quality as well as its colour and damaging its reputation. Trebbiano is now very much an optional ingredient, however, which is increasingly spurned by the quality-conscious red wine producers of Tuscany.

The sea of Trebbiano still produced in Tuscany is being diverted into cool-vinified innocuous dry whites such as Galestro. As Viura (MACABEO) displaced the more interesting local white grape variety MALVASIA in Rioja, so the high-yielding Trebbiano has largely ousted Malvasia from Tuscany, although the latter makes a much more serious base for vin santo.

Italy's most exciting Trebbiano, Valentini's Trebbiano d'Abruzzo, is probably not made from Trebbiano at all but BOMBINO BIANCO. Nevertheless, there was a dramatic increase in plantings of a variety known as Trebbiano d'Abruzzo in the 1980s.

Trebbiano has also managed to infiltrate Portugal's fiercely nationalistic vineyards, as Thalia, and is widely planted in Bulgaria and in parts of Greece and Russia. As well as being used for Mexico's important brandy production, Trebbiano is well entrenched in the southern hemisphere, where its high yields and high acidity are valued. There were more than 4,000 ha / 9,800 acres of 'Ugni Blanc' in Argentina by the end of the 1980s as well as extensive plantings in Brazil and Uruguay.

South Africa also calls its relatively limited plantings Ugni Blanc but relies more on COLOMBARD for brandy production and cheap, tart blending material, as does California, whose few hundred acres of remaining 'St-Emilion' are exclusively in the Central Valley. Australia, however, has twice as much Trebbiano as Colombard, planted mainly in the irrigated areas, where it provides a usefully tart ingredient in basic blended whites and is also sometimes used by distillers, and can occasionally be fashioned into a full bodied wine with substance, just as the odd Californian manages.

The influence of Trebbiano/Ugni Blanc will surely decline as wine drinkers seek flavour with increasing determination.

See also UGNI BLANC.

TREIXADURA, Galician name for Portugal's scented, delicate white TRAJADURA and treated in much the same way. This is the main grape of Ribeiro and may be blended with Galician TORRONTÉS and LADO.

TREPAT, indigenous speciality of the Conca de Barberá and Costers del Segre regions in north east Spain. About 1,500 ha/3,700 acres are used mainly for light rosés.

TRESALLIER, vine grown in France's Allier *département*, notably for St-Pourçain, that is closely related to the SACY of the Yonne. It is certainly traditional there, but not unanimously acclaimed nowadays.

TRESSOT, Burgundian name for Jura's TROUSSEAU.

TRINCADEIRA PRETA, widely used Portuguese synonym.

TRIOMPHE (D'ALSACE), hybrid bred by Kuhlmann in Alsace from KNIPPERLÉ and a *riparia-rupestris* American vine. It has good resistance to mildew but has a tell-tale aroma with traces of foxiness.

TROLLINGER, most common German name for the distinctly ordinary black grape known as SCHIAVA by Italian speakers and Vernatsch by German speakers in the Italian Tyrol, and Black Hamburg by many who grow and buy table grapes. It almost certainly originated in what is now the Italian Tyrol and its German name is a corruption of *Tirolinger*. In Germany it is grown almost exclusively in Württemberg, where it was known as early as the 14th century. Indeed only 10 of the 2,300 ha (5,700 acres) of Trollinger counted in German vineyards in 1991 grew outside this distinctive region of smallholdings centred on Stuttgart, whose speciality is red, or at least dark pink, wine. To suit local tastes, and most Trollinger is consumed within Württemberg and within the year, the wines are often relatively sweet and are rarely worth serious study. Growers coax absurdly high yields from this willing provider, the most productive clone of Vernatsch known as Schiava

Grossa or Grossvernatsch. Although the variety remains popular with growers in Württemberg, where it constitutes one vine in every three, its relatively late ripening (later than Riesling) makes it unpopular elsewhere in Germany.

TROUSSEAU, one of the two principal dark grapes of the Jura region in eastern France. It is more robust and deeply coloured than POULSARD although both are in serious decline, relative to PINOT NOIR. It buds late and therefore tends to avoid spring frost damage but is an irregular yielder. The 19th century French ampelographer Comte A. Odart noted that Trousseau is the same as Portugal's BASTARDO, and a variety for long called Cabernet Gros and occasionally, erroneously, Touriga in Australia. In both of these countries it is commonly used for sweet, dessert wines and flourishes in very different conditions from its native Jura. Galet also posits that the relatively rare variety known as Malvoisie Noire in the Lot may be Trousseau.

The lighter-berried mutation **Trousseau Gris** is the variety called Gray Riesling in California, where there were still nearly 300 acres / 120 ha planted in 1991, mainly in cooler sites. Just how this relatively obscure variety made its way around the world remains a mystery, although a number of other cuttings were taken from Portugal to Australia.

TSAOUSSI, speciality of the Greek island of Cephalonia, where it may be blended with the most distinctive ROBOLA.

TSOLIKAURI, relatively important wine grape in Georgia, although only about a tenth as widely planted there as the popular RKATSITELI.

U

UGNI BLANC (alias Italy's ubiquitous TREBBIANO), France's *éminence grise*. Although hardly ever seen on a wine label, it is France's most planted white grape by far, outnumbering Chardonnay's area five to one at the end of the 1980s. Like AIRÉN, Spain's most planted grape, it is used mainly for distilling into Spanish brandy, so the copious, thin, acid wine of Ugni Blanc washes through France's cognac and armagnac stills. It represents 95 per cent of all vines planted in the Cognac region, where it is often called St-Émilion. Armagnac is also made from BACO BLANC and COLOMBARD. Ugni Blanc supplanted the FOLLE BLANCHE that was the main ingredient in French brandy production before phylloxera arrived, because of its good resistance to powdery mildew and grey rot.

It was probably imported from Italy during the 14th century, when the papal court was established at Avignon. Other Italian varieties were presumably similarly transported but Trebbiano's extraordinarily high yields and high acidity may have helped establish it in southern France, where it is still grown widely today. Often called Clairette Ronde (although not related to CLAIRETTE) it was still among the five most planted varieties in the southern Rhône and Provence, in most of whose appellations it is allowed to play a subsidiary role. There are still small plantings of it in Corsica and it was only in the late 1980s that Sauvignon Blanc overtook Ugni Blanc as Bordeaux's second most planted white grape variety. It is still widely planted in the north west of the Gironde where, like Colombard, it is allowed into Bordeaux Blanc up to 30 per cent and is sometimes tellingly called Muscadet Aigre, or 'sour Muscadet'.

For more details of this variety, which probably produces more wine than any other, see TREBBIANO (although it is usually called Ugni Blanc in South America, where it is widely planted).

ULL DE LLEBRE, meaning 'hare's eye', Catalan name for TEMPRANILLO.

UVA ABRUZZI, occasional name for Italy's MONTEPULCIANO.

🍇 **UVA DI TROIA,** southern Italian grape fast declining in popularity with growers. The vine's connection with Troy remains a mystery. Italy's total plantings, mainly in Apulia, fell from 5,000 ha in 1982 to about 3,000 ha / 7,400 acres in 1990.

🍇 **UVA RARA,** variety too widely grown in Oltrepò Pavese in Lombardy in northern Italy to justify its Italian name, whose literal translation is 'rare grape'. Often called, misleadingly, Bonarda Novarese (see BONARDA), it is grown in the Novara hills and used to soften the SPANNA grapes grown here in a range of scented red wines.

V

🍇 **VACCARÈSE,** minor, large-berried southern Rhône variety officially but usually only theoretically allowed into Châteauneuf-du-Pape. It is grown at Château de Beaucastel, where it is regarded as a sort of cousin of CINSAUT.

VALDEPEÑAS, not particularly widely grown grape in California which may be Spain's TEMPRANILLO.

🍇 **VALDIGUIÉ,** sometimes called Gros Auxerrois, enjoyed its finest hour in the late 19th century, when it was valued for its productivity and its resistance to powdery mildew in the Lot, South West France. In the early 20th century it was known as 'the ARAMON of the south west' for its emphasis on quantity at the expense of quality and has been all but eradicated from France, where there remained only a few hundred ha by the 1988 census.

Eight years earlier French ampelographer Galet had visited the USA and identified the variety then sold rather successfully as Napa Gamay as none other than this undistinguished vine, of which there were then 4,000 acres/1,600 ha planted in California. By the 1990s it had all but disappeared from official statistics (although the giant Gallo company has toyed with the idea of releasing a varietal version).

VAVRINECKE, Czech name for ST LAURENT.

VELTLINER, VELTLIN ZELENE, VELTLINSKE ZELENÉ, VELTLINI, common synonyms for the Austrian white grape variety GRÜNER VELTLINER, also known in darker-skinned mutations as **Roter Veltliner,** which can produce characterful wines in ripe years, and occasionally **Brauner Veltliner.** Another important white grape variety, chiefly encountered in Austria, is FRÜHROTER VELTLINER, sometimes known as Veltliner in the far north east of Italy.

🍇 **VERDEA,** speciality of the Colli Piacenti, north central Italy.

🍇 **VERDECA,** Apulia's most popular light-berried vine producing neutral wine suitable for the vermouth industry and declining in popularity. Total plantings fell by 2,000 ha to 4,000 ha/10,000 acres between 1982 and 1990.

🍇 **VERDEJO,** characterful grape with distinctive blue-green bloom that is the Spanish Rueda region's pride and joy (and has staved off a challenge for primacy from imported SAUVIGNON BLANC, with which it is often blended). Wines produced are aromatic, herbaceous (with a scent somewhat reminiscent of laurel), but with great substance and extract, capable of ageing well into an almost nutty character. Also grown in Cigales, Toro.

🍇 **VERDELHO,** Portuguese grape most closely associated with the island of Madeira, where it became increasingly rare in the post-phylloxera era but the name was for long used to denote a style of wine somewhere between SERCIAL and BUAL levels of richness. The relatively few Verdelho vines on Madeira produce small, hard grapes and musts high in acidity. It is probably the same as the GOUVEIO of the Douro and possibly the same as the VERDELLO of Italy. The variety has had its most notable success in vibrant, lemony, substantial table wines in Australia, most notably in some of the hotter regions of Western Australia.

A dark-skinned grape called **Verdelho Tinto** is known on Madeira.

🍇 **VERDELLO,** Italian vine name known in both Umbria and Sicily. A connection with VERDELHO seems likely but is unproven.

🍇 **VERDESSE,** minor vine of Bugey in eastern France whose wine can be powerful and highly aromatic.

🍇 **VERDICCHIO,** central Italian vine, with many subvarieties, provider of classic varietals on the Adriatic coast. The best examples show lemony acidity and bitter almonds; the worst are just crisp dry whites. Total Verdicchio plantings were nearly 4,000 ha/10,000 acres in the early 1990s.

🍇 **VERDIL, VERDOSILLA,** south eastern Spaniard making rather neutral wine in Yecla and the southernmost part of Valencia. Fewer than 10 ha/25 acres remained in the mid 1990s.

🍇 **VERDISO,** lively speciality of Treviso in north east Italy.

🍇 **VERDONCHO** undistinguished La Mancha grape.

🍇 **VERDOT.** The variety known as Verdot in Chile, where there were a few hundred ha in 1991, is probably Bordeaux's PETIT VERDOT, but the Verdot of which there were 100 ha/250 acres in Argentina's Mendoza in 1989 may well be the coarser, probably unrelated, GROS VERDOT variety.

🍇 **VERDUZZO, VERDUZZO FRIULANO,** historic north east Italian grape grown mainly in Friuli (although not in Piave). The best comes from the hills of the Colli Orientali zone. The wine exists both in a dry and a sweet version, although the latter, obtained either by late harvesting or by raisining the grapes can frequently be more medium dry than lusciously sweet. Sweet Verduzzo is less common but more interesting than dry, being deep gold and often with a delightful density and honeyed aromas, even if it lacks the complexity of an outstanding dessert wine. Ramandolo is the classic sweet wine zone. The natural astringency of Verduzzo grapes is more noticeable in the dry version.

Italy's 1990 vineyard survey found 1,800 ha/4,500 acres of Verduzzo Friulano, and 2,600 ha of **Verduzzo Trevigiano**, the much less characterful grape grown in the Veneto to produce somewhat neutral dry whites.

🍇 **VERMENTINO,** attractive, aromatic grape widely grown in Sardinia, Liguria, to a limited extent in Corsica, and to an increasing extent in Languedoc-Roussillon, where it is a recently permitted variety in many appellations, including white Côtes du Roussillon. It is thought to be identical to the variety long grown in eastern Provence as ROLLE. In Corsica it is often called Malvoisie de Corse, and some believe that the variety is related to the MALVASIA family. Vermentino is Corsica's most planted white grape (although it is essentially a red wine-producing island) and dominates the island's white appellation contrôlée wines. There were only about 400 ha/1,000 acres planted on Corsica at the end of the 1980s.

Italy has nearly 4,000 ha of Vermentino vines mainly, but not exclusively, grown in Sardinia and in the Ligurian vineyards around Genoa. In Sardinia it is picked deliberately early to retain acid levels but still manages to produce lively wines of character. Body, acid, and perfume are its hallmarks, a good combination.

VERNACCIA, name used for several, unrelated Italian grape varieties, mainly white but sometimes red, from the extreme north of the peninsula (VERNATSCH being merely a Germanic version of Vernaccia) to the fizzy red **Vernaccia di Serrapetrona** of the Marches and the **Vernaccia di Oristano** which is an almost sherry-like wine made on the island of Sardinia. The most highly regarded form is VERNACCIA DI SAN GIMIGNANO, which has no relationship to the Sardinian **Vernaccia di Cagliari** vine, according to the studies of Professor Liuzzi of Cagliari in the early 1930s.

The name is so common because it comes from the same root as the word 'vernacular' or indigenous. Wines called Vernaccia, or sometimes Vernage, are often cited in the records of London wine merchants in the Middle Ages, but the term could have been used for virtually any sort of wine, Latin being the common language then. Vernaccia was a particularly common product of Liguria in north west Italy and Tuscany.

VERNACCIA DI SAN GIMIGNANO, VERNACCIA local to the sandstone-based vineyards around the many-towered Tuscan town of San Gimignano. There are references to Vernaccia in the archives of San Gimignano as early as 1276. Just after the Second World War TREBBIANO and MALVASIA threatened to take over the zone but Vernaccia's very positive, almost varnishy but crisp, full bodied quality has won it recent favour—along with attempts to add even more interest with barrel ageing.

A dark-berried **Vernaccia Nera**, sometimes called **Vernaccia di Serrapetrona**, survives on Italy's Adriatic coast.

VERNATSCH, German name for the undistinguished light red grape variety SCHIAVA.

VESPAIOLA, vine grown in the Veneto region of north east Italy, said to take its name from the wasps (*vespe*) attracted by the high sugar levels of its ripe grapes. Its most famous product is Maculan's Torcolato sweet wine of Breganze, although in this Vespaiola is blended with TOCAI FRIULANO and GARGANEGA.

VESPOLINA, vine native to the Gattinara region of northern Piedmont, north west Italy, where it is often blended with the more powerful NEBBIOLO. According to Anderson it is also grown, as Ughetta, in Lombardy.

🍇 VIDAL, French hybrid more properly known as **Vidal Blanc** or **Vidal 256** and widely grown in Canada, where it is particularly valued for its winter hardiness. Grown to a limited extent in the eastern United States, it is a hybrid of UGNI BLANC and one of the SEIBEL parents of SEYVAL BLANC. The wine produced, like Seyval's, has no obviously foxy character and can smell attractively of currant bushes or leaves. Its slow, steady ripening and thick skins make it particularly suitable for sweet, late harvest (non-botrytized) wines and Ice Wine, for which it, with RIESLING, is famous in Canada. Vidal-based wines do not have the longevity of fine Rieslings, however. Plantings of Vidal total about 1,000 ha/2,500 acres.

🍇 VIEN DE NUS, speciality of the town of Nus in Italy's Valle d'Aosta.

🍇 VILANA, speciality of the Greek island of Crete, where it can produce lively, quite delicate wines around Peza designed to be drunk young.

VILLARD, common French name for a great French viticultural secret, their most commonly planted hybrids of the vast SEYVE-VILLARD group, mainly, like England's widely planted SEYVAL BLANC, based on various members of the SEIBEL family. **Villard Noir** is Seyve-Villard 18.315, planted all over France, from the northern Rhône to Bordeaux, and was treasured for its resistance to downy mildew. **Villard Blanc**, or Seyve-Villard 12.375, made slightly more palatable wine (though the must can be difficult to process). Both varieties yield prodigiously and for that attribute were so beloved by growers that in 1968 there were 30,000 ha/74,000 acres of Villard Noir and 21,000 ha of Villard Blanc in France (making them fifth and third most planted black and white grape varieties respectively).

To the great credit of the authorities, and thanks to not inconsiderable bribes for grubbing up, within 20 years these totals had been shaved to 2,500 ha and 4,600 ha respectively and Villards of both hues will be almost completely eradicated by the turn of the century. The growers of Tarn and Ardèche are particularly attached to Villard Noir.

Other hybrids once widely planted in France are BACO, CHAMBOURCIN, COUDERC, PLANTET, and various other members of the Seibel and Seyve-Villard families.

🍇 VIOGNIER, highly fashionable grape since the early 1990s, partly because its most famous wine Condrieu is both distinctive and, most importantly, scarce.

The vines can withstand drought well but are prone to powdery mildew. The grapes are a deep yellow and the resulting wine is high in colour, alcohol, and a very particular perfume redolent of apricots, peaches, and blossom. Condrieu is one of the few highly priced white wines that should probably be drunk young, while this perfume is at its most heady and before the wine's slightly low acidity fades.

The vine was at one time a common crop on the farmland south of Lyon and has been grown on the infertile terraces of the northern Rhône for centuries but its extremely low productivity here saw it decline to an official total of just 14 ha/35 acres in the French agricultural census of 1968—mostly in the three north Rhône appellations in which it is allowed, Condrieu, Château Grillet, and, to an even lesser extent, the dense red Côte Rôtie, in which it is allowed as a perfuming agent up to 20 per cent of the SYRAH-dominated total.

French nurserymen saw an increase in demand for Viognier cuttings from the mid 1980s, however (when the red wines of the Rhône enjoyed a renaissance of popularity), and by 1988 were selling half a million a year. In the early 1990s the consumer could choose from a range of recognizably perfumed, if slightly light, southern French varietal Viogniers, some of them produced at quite high yields, from cuttings field-grafted on to less fashionable vine varieties. In the Languedoc-Roussillon there was not a single Viognier vine before 1989, but there were already 140 ha/350 acres by 1993, and many more now. It is often blended with other relative newcomers to the Languedoc-Roussillon such as MARSANNE and ROLLE. As well as being grown in the Ardèche and throughout the Languedoc-Roussillon, Viognier became increasingly popular in California because of its modish associations with the Rhône. All but a third of the state's modest total acreage of Viognier in 1992 was too young to bear fruit.

It is also increasingly planted in Australia and elsewhere. Galet also reports plantings of Viognier at Garibaldi in Brazil.

🍇 **VIOSINHO,** useful, crisp northern Portuguese grape grown in the Douro and Trás-os-Montes.

🍇 **VITAL,** western Portuguese grape producing slightly flabby, undistinguished wines. Known as Malvasia Corado in the north of the country, according to Mayson.

VIURA is a common Spanish synonym for MACABEO and is therefore what Riojanos call their dominant white grape variety.

VLASSKY RIZLING, Czech name for WELSCHRIESLING.

🍇 **VOLIDZA,** exciting Greek rarity from the same area as MAVROUDI.

🍇 **VRANAC,** powerful speciality of Montenegro in what was Yugoslavia. The wines produced are deep in colour and can be rich in extract, responding unusually well to oak ageing. There is an element of refreshing bitterness on the finish of these wines that suggests some relationship with an Italian variety just across the Adriatic. Vranac is one of the few indigenous grape variety names to appear on the label of wines exported from what was ugoslavia. It is also grown, less successfully, in Macedonia and plantings total only a few hundred acres.

W

WÄLSCHER, occasional Austrian synonym for CHASSELAS.

WÄLSCHRIESLING. See WELSCHRIESLING.

WALTHAM CROSS, occasional name for the table grape DATTIER.

WEIHRAUCHTRAUBE, or 'frankincense grape', is the German synonym for TĂMÎIOASĂ ROMÂNEASCĂ.

WEISS or **WEISSER**, adjective meaning 'white' in German, often used for light-berried vine varieties.

WEISSBURGUNDER, WEISSERBURGUNDER, WIESSER BURGUNDER, common German synonyms for PINOT BLANC, somewhat misleading in associating the variety specifically with Burgundy, where it is relatively unimportant today.

WEISSER RIESLING, common synonym for the great white RIESLING grape variety of Germany.

WELDRA, South African crossing of CHENIN BLANC × TREBBIANO with good acid levels but not much flavour. See also CHENEL.

WELSCHRIESLING, or **WÄLSCHRIESLING**, important central European vine which, as Germans are keen to point out, is completely unrelated to the great RIESLING grape of Germany. Indeed it rankles many Germans that the noble word is even allowed as a suffix in its name; they would prefer that the word Rizling were used, as in **Welsch Rizling** or **Welschrizling**, which it is in many of its many synonyms.

Welschriesling may be the variety's most common name in Austria, but **Welschrizling** is obediently used in Bulgaria; its most common name in Hungary (where it is the country's most planted grape) is OLASZ RIZLING; in Slovenia and Vojvodina it is LASKI RIZLING (it is also the most planted grape in former Yugoslavia); and in what was Czechoslovakia it is the very similar Rizling Vlassky. Only in Croatia does it acquire a name of any distinction, Graš evina. The Italians call it Riesling Italico (as opposed to Riesling Renano, which is the Riesling of Germany) and variants of this are used all over eastern Europe, notably in Romania (although most of the 'Riesling' planted in the ex-Soviet republics is true German Riesling). The variety is one of the few common white wine grapes in Albania, as it is in its close political ally China.

The origins of this old variety are obscure. Although French origins have been posited, this seems unlikely as it is quite

199

unknown in France (and Germany). It thrives best in dry climates and warmer soils, and has a tendency to produce excessively acid wines in cool climates.

Welsch simply means 'foreign' in Germanic languages, which provides few clues. But since Vlaska is the Slav name for Wallachia in Romania, and since the variety is particularly successful in that country, it is easy to develop a theory that it originated there, and that Laski is a corruption of Vlassy, or Wallachian.

Although Welschriesling has little in common with Riesling, it too is a late ripening vine whose grapes keep their acidity well and produce light bodied, relatively aromatic wines. Welschriesling can easily be persuaded to yield even more productively than Riesling, however, and indeed this and its useful acidity probably explain why it is so widely planted throughout eastern Europe and, partly, why so much of the wine it produces has been undistinguished (although low technological standards in many eastern bloc wineries have also played a part, notably in the Slovenian export success Lutomer Laski Riesling of not so blessed memory to many of us).

As a wine Welschriesling reaches its apogee in Austria, specifically in some particularly finely balanced rich late harvest wines made on the shores of the Neusiedlersee in Burgenland, where about two-thirds of Austria's total 5,000 ha / 12,500 acres of the variety are planted. In particularly favoured vintages, and especially around the Neusiedlersee, noble rot forms to ripen grapes up to Trockenbeerenauslese level, while retaining the acidity that is Welschriesling's hallmark. Some producers use the grape to add finesse to Chardonnay. Welschriesling may not have the aromatic character of Germany's Riesling, but since aroma plays only a small part in the appreciation of really sweet wines, this leaves Welschriesling at less of a disadvantage than Riesling addicts might imagine, although Austrian Trockenbeerenauslese wines rarely have the longevity of their German counterparts. The bulk of Austria's Welschriesling, however, goes into light dryish wines for early drinking, notably in Burgenland and Styria. It has also been known as Riesler in Austria.

Riesling Italico [*sic*] was Romania's third most planted variety in the early 1990s, and was often blended with other varieties such as Muscat Ottonel. Within Italy it is most common in the far north east, in Friuli just over the border from Slovenia. Provided its tendency to overcrop is curbed, it can produce delicate, crisp, mildly flowery wines, most successfully in Collio. It is grown to a limited extent in Alto Adige and, more successfully, in Lombardy. In 1990 there were about 2,400 ha / 6,000 acres of Riesling Italico in Italy, while total plantings of true Riesling were less than 1,000 ha.

WHITE FRENCH, occasional South African name for PALOMINO.

WHITE RIESLING, common synonym for the great white RIESLING grape variety of Germany.

🍇 **WILDBACHER,** or **BLAUER WILDBACHER,** speciality of western Styria in southern Austria, where almost all of the 230 ha/570 acres grown in the early 1990s were located. The variety has been increasingly popular with growers and almost all of it is made into the local pink speciality, Schilcher wine, enlivened by Wildbacher's high acidity and distinctive perfume.

🍇 **WÜRZER,** Gewürztraminer × Müller-Thurgau crossing made at the German viticultural station of Alzey in 1932 and only planted in any significant quantity in the 1980s, at the end of which there were just over 100 ha/250 acres, mainly in Rheinhessen. It is overpoweringly heady, yields well, but a little goes a very long way indeed.

X

XANTE, XANTE CURRANT, see CURRANT.

🍇 **XAREL-LO,** Catalan curiosity important in the production of sparkling Cava, to which it can add an almost cabbagey, definitely vegetable-like, note. It is particularly important in Alella, where it is known as Pansa Blanca. It is most commonly found in Penedès, however, where, with PARELLADA and MACABEO, it makes up most Cava blends. The vine is very vigorous and productive and buds early, so is prone to spring frost damage. It needs careful pruning and the wine it produces can be very strongly flavoured.

🍇 **XYNISTERI,** the most common light-berried vine grown on the island of Cyprus and preferred to the dark-skinned MAVRO for the rich dessert wine Commandaria, which is the island's claim to wine fame.

🍇 **XYNOMAVRO,** most common Greek red wine grape grown all over northern Greece as far south as the foothills of Mount Olympus, where Rapsani is produced. Its name means 'acid black' and the wines can indeed seem harsh in youth but they age well, as mature examples of Naoussa can demonstrate. One of the few Greek vine varieties which may not reach full ripeness in some years, it is blended with a small proportion of the local NEGOSKA to produce Goumenissa and is also used as a base for sparkling wine on the exceptionally cool, high vineyards of Amindeo. Wines tend to be relatively soft but to have good acid and attractive bite.

Z

🍇 **ZALA GYÖNGYE,** Muscat-like table grape crossing of an Eger grape and Pearl of Csaba quite widely planted in Hungary, where some undistinguished wine is also made from it. It is also grown in Italy, Croatia, Romania, Israel, and elsewhere for table grapes and is usually called by a local translation of the expression 'Queen of the Vineyards'.

🍇 **ZALEMA,** Spanish grape grown particularly in the southern Condado de Huelva zone, where its musts and wine can oxidize easily. It is being replaced by higher-quality varieties such as PALOMINO.

🍇 **ZEFIR,** early-ripening 1983 Hungarian crossing of LEÁNYKA and HÁRSLEVELŰ producing soft, spicy wine.

🍇 **ZENIT,** 1951 Hungarian crossing of BOUVIER and EZERJÓ which ripens usefully early to produce crisp, fruity but not particularly aromatic wines.

ZIBIBBO, Sicilian name for MUSCAT OF ALEXANDRIA sometimes made into wine, notably the sweet orange Moscato di Pantelleria and some dried grape table wines from that island, although more usually sold as table grapes. The 1990 Italian vineyard survey found scarcely 1,800 ha / 4,500 acres of Zibibbo, as compared with 13,500 ha of Moscato Bianco (MUSCAT BLANC À PETITS GRAINS).

🍇 **ZIERFANDLER,** the more noble of the two white wine grape varieties traditionally associated with Gumpoldskirchen, the dramatically full bodied, long-lived spicy white wine of the Thermenregion district of Austria. (The other is ROTGIPFLER.) At the end of the 1980s there were about 120 ha / 300 acres of Zierfandler, which ripens very late, as its synonym Spätrot suggests, but keeps its acidity better than Rotgipfler. Unblended, Zierfandler has sufficient nerve to make late harvest wines with the ability to evolve over years in bottle, but most Zierfandler grapes are blended, and sometimes vinified, with Rotgipfler. The variety, as Cirfandli, is also known in Hungary.

🍇 **ZILAVKA,** relatively successful Bosnian variety in what was Yugoslavia. The variety manages to combine high alcohol with high acidity and a certain nuttiness of flavour. The Zilavka made around the inland town of Mostar has been particularly prized. It is also increasingly planted in Macedonia.

🍇 **ZINFANDEL,** exotic black-skinned grape grown predominantly in California that has tended to mirror the giddily changing fashions of the American wine business.

California historian Charles Sullivan points out that Zinfandel, or at least a vine called 'Zinfindal', was well known on the American East Coast in the first half of the 19th century. It was exhibited at the Massachusetts Horticultural Society as early as 1834 and is frequently mentioned in the agricultural press of the 1840s and 1850s. In 1858 a Sacramento nurseryman exhibited a grape variety he called 'Zeinfindall' at the California state fair and by 1860 several California wines had been made from it. It seems highly likely that these cuttings came from the East Coast, although one San Jose nurseryman claimed, retrospectively and unverifiably, to have imported his Zinfandel directly from France to California in 1852 as 'Black St Peters'.

Because Zinfandel has no French connection, it has escaped the detailed scrutiny of the world's ampelographic centre in Montpellier and its European origins rested on local hypothesis rather than internationally accredited fact until the application of DNA testing techniques to vines in the early 1990s. Only then was it irrefutably demonstrated that Zinfandel is one and the same as the PRIMITIVO of Apulia on the heel of Italy. The relationship had already been sufficiently acknowledged by the Italians in the 1980s that some were exporting their Primitivo labelled, in direct appeal to the American market, Zinfandel. There may also be a link with the PLAVAC MALI of Dalmatia. Some maintain that the vine was known in California before it was known in Apulia.

Zinfandel took firm hold on the California wine business in the 1880s, when its ability to produce in quantity was prized above all else. Many was the miner, and other beneficiary of California's gold rush, whose customary drink was Zinfandel. By the turn of the century Zinfandel was regarded as California's own claret and occupied some of the choicest North Coast vineyard. During Prohibition it was the choice of many a home wine-maker but since then its viticultural popularity has been its undoing.

In 20th century California Zinfandel has occupied much the same place as SHIRAZ (Syrah) in Australia and has suffered the same lack of respect simply because it is the most common black grape variety, often planted in unsuitably hot sites and expected to yield more than is good for it. Zinfandel may not be quite such a potentially noble grape variety as Syrah but it is certainly capable of producing fine wine if yields are restricted and the weather cool enough to allow a reasonably long growing season.

Zinfandel's viticultural disadvantages are that bunches can ripen unevenly, leaving harsh, green berries on the same bunch as those that have reached full maturity, and that once grapes reach full ripeness, in direct contrast to its great California rival Cabernet Sauvignon for example, they will soon turn to raisins if not picked quite rapidly. It therefore performs best in warm but not hot conditions and high-altitude vineyards can work particularly well.

Although Zinfandel has been required to transform itself into virtually every style and colour of wine that exists, it is probably best suited to dry, sturdy, unsubtle but vigorous reds with an optimum lifespan of four to eight years. Such wines are rarely blends, although Ridge Vineyards are exponents of blending some PETITE SIRAH into Zinfandel. Dry Creek Valley in Sonoma has demonstrated a particular aptitude for this underestimated variety.

In the mid and late 1980s, the state's surplus of Zinfandel grapes was harnessed to the American public's apparently insatiable demand for crisp, light, fruity wine and a product called White Zinfandel (actually a very pale pink, made from Zinfandel grapes left in contact with the skins for only a very short time) was a tearaway commercial success.

Thanks to the enormous popularity of White Zinfandel, Zinfandel plantings, which had been declining, increased by up to 3,000 acres a year during the late 1980s, mostly in the Central Valley, so that they totalled nearly 39,000 acres / 15,800 ha in 1994, just ahead of California's total acreage of its second most important black grape variety CABERNET SAUVIGNON.

It is also grown to a much more limited extent in warmer sites in other western states, and some South African growers have taken advantage of their climate's suitability for Zinfandel. Australia is another obvious location for this unusual variety and Cape Mentelle in Western Australia has been particularly successful with it.

♀ **ZWEIGELT, ZWEIGELTREBE,** or **BLAUER ZWEIGELT,** Austria's most planted dark-berried grape, even though this crossing was bred only relatively recently, by a Dr Zweigelt at the Klosterneuburg research station in 1922. It is a BLAUFRÄNKISCH × ST LAURENT crossing that at its best combines some of the bite of the first with the body of the second, although it is sometimes encouraged to produce too much dilute wine. It is popular with growers because it ripens earlier than Blaufränkisch but buds rather later than St Laurent, thereby tending to yield generously. It is widely grown throughout all Austrian wine regions and can occasionally make a serious, age-worthy wine even though most examples are best drunk young. So successful has it been in Austria that the variety has also been planted on an experimental basis in Germany and England. The export fortunes of the variety may, oddly enough, be hampered by its originator's uncompromisingly Germanic surname. If only he had been called Dr Pinot Noir.

THE GRAPES BEHIND THE NAMES

It is still impossible to tell from the labels of most geographically named wines which grapes were used to make them. The following is a unique guide to the varieties officially allowed into the world's wine appellations, listed by local name alphabetically by country, grouped where appropriate into alphabetically listed regions within that country, and listed alphabetically by appellation name within that region.

Italics denote minor grapes. R, W, P, and S denote red, white, pink, and sparkling wines respectively.

AUSTRIA

Gumpoldskirchner (W) Rotgipfler, Zierfandler

Steiermark Schilcher (P) Blauer Wildbacher

CROATIA

Hvar (W) Trbljan
Pelješac, Postup, Dingač, Prošek, Faros, Potomje (R) Plavac Mali

CYPRUS

Commandaria (R) Mavro, Xynisteri

FRANCE

ALSACE AND NORTH EAST

Crémant d'Alsace (S) Riesling, Pinot Blanc, Pinot Noir, Pinot Gris, Auxerrois, Chardonnay

Vin d'Alsace Edelzwicker (W) Gewurztraminer, Riesling, Pinot Gris, Muscat Blanc à Petits Grains, Muscat Ottonel, Pinot Blanc, Pinot Noir, Sylvaner, Chasselas
Other Alsace wines are varietally labelled

Côtes de Toul (R) Pinot Meunier, Pinot Noir   (P) Gamay, Pinot Meunier, Pinot Noir, *Aligoté, Aubin, Auxerrois*   (W) Aligoté, Aubin, Auxerrois

Vins de Moselle Auxerrois, Gewürztraminer, Pinot Meunier, Müller-Thurgau, Pinot Noir, Pinot Blanc, Pinot Gris, Riesling, *Gamay*

BORDEAUX

Blaye (R) Cabernet Sauvignon, Cabernet Franc, Merlot, Malbec, *Prolongeau (Bouchalès), Béquignol, Petit Verdot*   (W) Merlot Blanc, Folle Blanche, Colombard, *Pineau de la Loire (Chenin Blanc), Frontignan, Sémillon, Sauvignon Blanc, Muscadelle*

Bordeaux, Bordeaux Clairet, Merlot, Bordeaux Supérieur (R, P) Cabernet Sauvignon, Cabernet Franc, Carmenère, *Malbec, Petit Verdot*

The Grapes Behind the Names

Bordeaux Sec (W) Sémillon, Sauvignon Blanc, Muscadelle, *Merlot Blanc, Colombard, Mauzac, Ondenc, Ugni Blanc*

Bordeaux Côtes de Francs (R) Cabernet Sauvignon, Cabernet Franc, Merlot, *Malbec* (W) Sémillon, Sauvignon Blanc, *Muscadelle*

Bordeaux Haut-Benauge (W) Sémillon, Sauvignon Blanc, *Muscadelle*

Bordeaux Mousseux (R, S) Cabernet Sauvignon, Cabernet Franc, Merlot, *Carmenère, Malbec, Petit Verdot* (W, S) Sémillon, Sauvignon Blanc, Muscadelle, *Ugni Blanc, Merlot Blanc, Colombard, Mauzac, Ondenc*

Bourg, Côtes de Bourg, Bourgeais (R) Cabernet Sauvignon, Cabernet Franc, Merlot, Malbec, *Gros Verdot, Prolongeau (Bouchalès)* (W) Sauvignon Blanc, Sémillon, *Muscadelle, Merlot Blanc, Colombard*

Côtes de Blaye (W) Sémillon, Sauvignon Blanc, Muscadelle, *Merlot Blanc, Folle Blanche, Colombard, Pineau de la Loire (Chenin Blanc)*

Côtes de Castillon (R) Cabernet Sauvignon, Cabernet Franc, Merlot, *Malbec (Côt)*

Crémant de Bordeaux (R, S) Cabernet Sauvignon, Cabernet Franc, Merlot, *Carmenère, Malbec, Petit Verdot, Ugni Blanc, Colombard* (W, S) Sémillon, Sauvignon Blanc, Muscadelle, *Ugni Blanc, Colombard* (P, S) Cabernet Sauvignon, Cabernet Franc, Merlot, *Carmenère, Malbec, Petit Verdot*

Entre-Deux-Mers, Entre-Deux-Mers Haut-Benauge (W) Sémillon, Sauvignon Blanc, Muscadelle, *Merlot Blanc, Colombard, Mauzac, Ugni Blanc*

Fronsac, Canon Fronsac, Côtes Canon Fronsac (R) Cabernet Sauvignon, Cabernet Franc (Bouchet), Merlot, *Malbec (Pressac)*

Graves (R) Cabernet Sauvignon, Cabernet Franc, Merlot, *Malbec, Petit Verdot*

Graves Supérieures (W) Sémillon, Sauvignon Blanc, Muscadelle

Graves de Vayres (R) Cabernet Sauvignon, Cabernet Franc, Merlot, *Petit Verdot* (W) Sémillon, Sauvignon Blanc, Muscadelle, *Merlot Blanc*

Margaux, Listrac-Médoc, Pessac-Léognan, Premières Côtes de Bordeaux (R) Cabernet Sauvignon, Cabernet Franc, Merlot, *Carmenère, Malbec (Côt), Petit Verdot*

Pauillac, St-Estèphe, St-Julien, Médoc, Haut-Médoc, Moulis (R) Cabernet Sauvignon, Cabernet Franc, Merlot, *Carmenère, Malbec, Petit Verdot, Gros Verdot*

Pomerol, Lalande-de-Pomerol, Néac, Lussac-St-Émilion, Montagne-St-Émilion, Parsac-St-Émilion, Puisseguin-St-Émilion, St-Georges-St-Émilion (R) Cabernet Sauvignon, Cabernet Franc (Bouchet), Merlot, *Malbec (Pressac)*

Premières Côtes de Blaye (R) Cabernet Sauvignon, Cabernet Franc, Merlot, *Côt* (W) Sémillon, Sauvignon Blanc, Muscadelle, *Merlot Blanc, Colombard, Ugni Blanc*

St-Émilion (R) Merlot, Cabernet Sauvignon, Cabernet Franc, *Carmenère, Malbec (Côt)*

Ste-Foy-Bordeaux (R) Cabernet Sauvignon, Cabernet Franc, Malbec, *Petit Verdot* (W) Sémillon, Sauvignon Blanc, Muscadelle, *Merlot Blanc, Colombard, Mauzac, Ugni Blanc*

Sauternes, Barsac, Ste-Croix-du-Mont, Loupiac, Cadillac, Cérons, Premières Côtes de Bordeaux, Pessac-Léognan, Côtes de Bordeaux, St-Macaire (W) Sémillon, Sauvignon Blanc, *Muscadelle*

BURGUNDY

REGIONAL APELLATIONS

Bourgogne Grand Ordinaire (R) Pinot Noir, Gamay, *César, Tressot* (P) *Pinot Gris, Pinot Blanc, Chardonnay* (W) Chardonnay, Aligoté, Pinot Blanc, Melon de Bourgogne, *Sacy*

Bourgogne Passe-tout-grains (R) Gamay, *Pinot Noir, Pinot Blanc, Pinot Gris, Chardonnay*

Bourgogne Mousseux (S) Chardonnay, Pinot Blanc, Pinot Noir, Pinot Beurot, Pinot Liébault, *César, Tressot*

Bourgogne, Bourgogne Clairet, Bourgogne Rosé, Bourgogne Hautes Côtes de Beaune, Bourgogne Rosé Hautes Côtes de Beaune, Bourgogne Clairet Hautes Côtes de Beaune, Bourgogne Irancy, Bourgogne Hautes Côtes de Nuits, Bourgogne Rosé Hautes Côtes de Nuits, Bourgogne Clairet Hautes Côtes de Nuits, Bourgogne Côte Chalonnaise, Bourgogne Rosé Côte Chalonnaise, Bourgogne Clairet Côte Chalonnaise (R, P) Pinot Noir, Pinot Liébault, *César, Tressot, Pinot Blanc, Pinot Gris, Chardonnay* (W) Chardonnay, *Pinot Blanc*

Bourgogne Aligoté, Bourgogne Aligoté Bouzeron (W) Aligoté, *Chardonnay*

Crémant de Bourgogne (S) Pinot Noir, Chardonnay, Pinot Gris, Pinot Blanc, *Gamay, Aligoté, Melon de Bourgogne, Sacy*

BEAUJOLAIS

Beaujolais, Beaujolais Supérieur, Beaujolais Villages (R, P) Gamay, *Pinot Noir, Pinot Gris, Chardonnay, Aligoté, Melon de Bourgogne* (W) Chardonnay, Aligoté

Brouilly, Chénas, Chiroubles, Fleurie, Juliénas, Morgon, Moulin-à-Vent, St Amour (R) Gamay, *Chardonnay, Aligoté, Melon de Bourgogne*

Côte de Brouilly (R) Gamay, *Pinot Noir, Pinot Gris, Pinot Blanc, Chardonnay*

Régnié (R) Gamay

CHABLIS

Petit Chablis, Chablis, Chablis Premier Cru, Chablis Grand Cru (W) Chardonnay

CÔTE CHALONNAISE

Givry (R) Pinot Noir, *Pinot Beurot, Pinot Liébault, Chardonnay* (W) Chardonnay, *Pinot Blanc*

THE GRAPES BEHIND THE NAMES

Mercurey*, Rully* (R) Pinot Noir, *Pinot Beurot, Pinot Liébault, Chardonnay* (W) Chardonnay

Montagny* (W) Chardonnay

CÔTE D'OR

Note: The communes indicated with an asterisk contain Premier Cru vineyards.

Côte de Beaune, Côte de Beaune-Villages, Côte-de-Nuits-Villages, Auxey-Duresses*, Beaune*, Blagny, Chassagne-Montrachet*, Chorey-lès-Beaune, Fixin*, Ladoix, Meursault*, Monthélie, Nuits-St-Georges*, Pernand-Vergelesses*, Puligny-Montrachet*, St-Aubin*, St-Romain, Santenay*, Savigny-lès-Beaune*, Vins Fins de la Côte de Nuits, Vougeot* (R) Pinot Noir, *Pinot Liébault, Chardonnay, Pinot Blanc, Pinot Gris* (W) Chardonnay, *Pinot Blanc*

Aloxe Corton*, Maranges*, Morey St-Denis* (R) Pinot Noir, *Pinot Liébault, Chardonnay, Pinot Blanc, Pinot Gris* (W) Chardonnay

Chambolle-Musigny*, Gevrey-Chambertin*, Pommard*, Volnay*, Volnay Santenots*, Vosne-Romanée* (R) Pinot Noir, *Pinot Liébault, Chardonnay, Pinot Blanc, Pinot Gris*

Marsannay (R) Pinot Noir, *Pinot Gris, Chardonnay* (W) Chardonnay, *Pinot Blanc*

Marsannay Rosé (P) Pinot Noir, Pinot Gris

CHABLIS GRAND CRU (W) See Chablis

CÔTE D'OR—GRANDS CRUS

Bonnes-Mares, Chambertin, Chambertin-Clos-de-Bèze, Chapelle-Chambertin, Charmes-Chambertin, Griotte-Chambertin, Latricières-Chambertin, Mazis-Chambertin, Mazoyères-Chambertin, Ruchottes-Chambertin, Clos des Lambrays, Clos de la Roche, Clos St-Denis, Clos de Tart, Clos de Vougeot, Echézeaux, Grands Echézeaux, La Grande Rue, Richebourg, Romanée-Conti, Romanée-St-Vivant, La Romanée, La Tâche, La Grande Rue (R) Pinot Noir, *Pinot Liébault, Pinot Blanc, Pinot Gris, Chardonnay*

Musigny, Corton (R) Pinot Noir, *Pinot Liébault, Pinot Blanc, Pinot Gris, Chardonnay* (W) Chardonnay

Corton-Charlemagne, Montrachet, Bâtard-Montrachet, Bienvenues-Bâtard-Montrachet, Chevalier-Montrachet, Criots-Bâtard-Montrachet (W) Chardonnay

Charlemagne (W) Chardonnay, *Aligoté*

MÂCONNAIS

Mâcon, Mâcon Supérieur, Mâcon-Villages, or Mâcon followed by a commune name (e.g. Mâcon Chardonnay, Mâcon Viré, Mâcon Lugny) (W) Chardonnay, *Pinot Blanc*

Mâcon, Mâcon Supérieur, or Mâcon followed by a commune name (e.g. Mâcon Chardonnay, Mâcon Viré, Mâcon Lugny) (R, P) Gamay, Pinot Noir, Pinot Gris

Pouilly-Fuissé, Pouilly-Vinzelles, Pouilly-Loché, St-Véran (W) Chardonnay

CHAMPAGNE

Champagne, Coteaux Champenois (S) Pinot Noir, Pinot Meunier, Chardonnay

Rosé de Riceys (P) Pinot Noir

CORSICA See Provence.

LANGUEDOC-ROUSSILLON

Cabardès, Côtes du Cabardès VDQS (R, P) Grenache, Syrah, Cinsaut, *Cabernet Sauvignon, Merlot, Cabernet Franc, Côt, Fer, Carignan, Aubun Noir*

Clairette de Bellegarde, Clairette du Languedoc (W) Clairette

Collioure (R) Grenache, Mourvèdre, *Carignan, Cinsaut, Syrah* (P) Grenache, Mourvèdre, *Carignan, Cinsaut, Syrah, Grenache Gris*

Corbières (R, P) Carignan, Grenache, Cinsaut, *Lladoner Pelut, Mourvèdre, Piquepoul Noir, Terret Noir, Syrah, Maccabeu, Bourboulenc, Grenache Gris* (W) Bourboulenc, Clairette Blanche, Grenache Blanc, Maccabeu, Muscat Blanc à Petits Grains, Piquepoul Blanc, Terret Blanc, Marsanne, Roussanne, Vermentino Blanc

Costières de Nîmes (R, P) Carignan, Grenache, Mourvèdre, Syrah, Cinsaut (W) Clairette Blanc, Grenache Blanc, Bourboulenc Blanc, Ugni Blanc

Coteaux du Languedoc (R) Carignan, Grenache, Lladoner Pelut, *Counoise Noir (Aubun), Grenache Rosé, Terret Noir, Picpoul Noir* (P) Carignan, Grenache, Lladoner Pelut, *Counoise Noir (Aubun), Grenache Rosé, Terret Noir, Picpoul Noir, Bourboulenc, Carignan Blanc, Clairette, Maccabéo, Picpoul, Terret, Ugni Blanc*

Côtes de la Malepère, VDQS (R) Merlot, Côt, Cinsaut, *Cabernet Sauvignon, Cabernet Franc, Grenache, Lladoner Pelut Noir, Syrah* (P) Cinsaut, Grenache, Lladoner Pelut Noir, *Merlot, Côt, Cabernet Sauvignon, Cabernet Franc, Syrah*

Côtes de Millau, VDQS (R) Gamay, Syrah, *Cabernet Sauvignon, Fer Servadou, Duras* (P) Gamay, *Syrah, Cabernet Sauvignon, Fer Servadou, Duras* (W) Chenin Blanc, Mauzac

Côtes du Roussillon (R, P) Carignan, Cinsaut, Grenache, Lladoner Pelut Noir, Syrah, Mourvèdre, Maccabeu Blanc (W) Grenache Blanc, Maccabeu Blanc, Tourbat Blanc/Malvoisie du Roussillon, Marsanne, Roussanne, Vermentino

Côtes du Roussillon Villages (R) Carignan, Cinsaut, Grenache, Lladoner Pelut Noir, Syrah, Mourvèdre, Maccabéo

Faugères (R, P) Carignan, Cinsaut, Grenache, Mourvèdre, Syrah, Lladoner Pelut Noir

Fitou (R) Carignan, Grenache, Mourvèdre, Syrah, Lladoner Pelut Noir, *Cinsaut, Maccabéo Blanc, Terret Noir*

Minervois (R, P) Grenache, Syrah, Mourvèdre, Lladoner Pelut Noir, Carignan, *Cinsaut, Picpoul Noir, Terret Noir, Aspiran Noir*   (W) Grenache Blanc, Bourboulenc Blanc (Malvoisie), Maccabeu Blanc, Marsanne Blanche, Roussanne Blanche, Vermentino Blanc (Rolle), *Picpoul Blanc, Clairette Blanche, Terret Blanc, Muscat Blanc à Petits Grains*

St-Chinian (R, P) Carignan, Cinsaut, Grenache, Lladoner Pelut Noir, Mourvèdre, Syrah

LOIRE AND CENTRAL FRANCE

Anjou, Anjou Gamay, Anjou pétillant, Rosé d'Anjou, Rosé d'Anjou pétillant, Saumur, Saumur Champigny, Saumur pétillant (R) Cabernet Franc, Cabernet Sauvignon, Pineau d'Aunis   (P) Cabernet Franc, Cabernet Sauvignon, Pineau d'Aunis, Gamay, Côt, Groslot   (W) Chenin Blanc, *Chardonnay, Sauvignon Blanc*

Anjou Coteaux de la Loire (W) Chenin Blanc

Anjou Mousseux (W, S) Chenin Blanc, *Cabernet Sauvignon, Cabernet Franc, Côt, Gamay, Groslot, Pineau d'Aunis*   (P, S) Cabernet Sauvignon, Cabernet Franc, Côt, Gamay, Groslot, Pineau d'Aunis

Anjou Villages (R) Cabernet Sauvignon, Cabernet Franc

Bourgueil, St Nicolas-de-Bourgueil (R) Cabernet Franc, *Cabernet Sauvignon*

Bonnezeaux (W) Chenin Blanc

Cabernet d'Anjou, Cabernet de Saumur (P) Cabernet Sauvignon, Cabernet Franc

Châteaumeillant (R, W, P) Gamay, Pinot Gris, Pinot Noir

Cheverny (R) Gamay, Pinot Noir, *Cabernet Franc, Cabernet Sauvignon, Côt*   (P) Gamay, Pinot Noir, *Cabernet Franc, Cabernet Sauvignon, Côt, Pineau d'Aunis*   (W) Sauvignon Blanc, *Chardonnay, Arbois (Menu Pineau), Chenin Blanc*

Chinon (W) Chenin Blanc (Pineau de la Loire)   (R, P) Cabernet Franc (Breton), *Cabernet Sauvignon*

Coteaux d'Ancenis (W) Chenin Blanc, Pinot Gris (Malvoisie)   (R, P) Cabernet Sauvignon, Cabernet Franc, Gamay, *Gamay de Chaudenay, Gamay de Bouze*

Coteaux de l'Aubance (W) Chenin Blanc (Pineau de la Loire)

Coteaux du Layon, Coteaux du Layon Chaume (W) Chenin Blanc (Pineau de la Loire)

Coteaux du Loir (W) Chenin Blanc (Pineau de la Loire)   (R) Pineau d'Aunis, Cabernet Franc, Cabernet Sauvignon, Gamay, Côt   (P) Pineau d'Aunis, Cabernet Franc, Cabernet Sauvignon, Gamay, Côt, Groslot

Coteaux de Saumur (W) Chenin Blanc (Pineau de la Loire)

Coteaux du Vendomois (W) Chenin Blanc, *Chardonnay*   (R) Pineau d'Aunis, *Gamay, Pinot Noir, Cabernet Franc, Cabernet Sauvignon*   (P) Pineau d'Aunis, *Gamay*

Côtes d'Auvergne (R, P) Gamay, Pinot Noir   (W) Chardonnay

Côtes de Gien, Coteaux du Giennois (R, P) Gamay, Pinot Noir   (W) Sauvignon Blanc

Cour-Cheverny (W) Romorantin Blanc

Crémant de Loire (W, P, S) Chenin Blanc, Cabernet Franc, Cabernet Sauvignon, Pineau d'Aunis, Pinot Noir, Chardonnay, Menu Pineau, *Grolleau Noir, Grolleau Gris*

Fiefs Vendéens (R, P) Gamay, Pinot Noir, *Cabernet Franc, Cabernet Sauvignon, Négrette, Gamay Chaudenay*   (W) Chenin Blanc, *Sauvignon Blanc, Chardonnay*

Gros Plant du Pays Nantais (W) Gros Plant (Folle Blanche)

Haut Poitou (W) Sauvignon Blanc, Chardonnay, Chenin Blanc, *Pinot Blanc* (R, P) Pinot Noir, Gamay, Merlot, Côt, Cabernet Franc, Cabernet Sauvignon, *Gamay de Chaudenay, Grolleau*

Jasnières (W) Chenin Blanc (Pineau de la Loire)

Menetou-Salon (W) Sauvignon Blanc (R, P) Pinot Noir

Montlouis (W) Chenin Blanc (Pineau de la Loire)

Muscadet, Musacadet Côtes de Grand Lieu, Muscadet des Coteaux de la Loire, Muscadet de Sevre et Maine (W) Melon

Pouilly-Fumé, Blanc Fumé de Pouilly (W) Blanc Fumé (Sauvignon Blanc)

Pouilly-sur-Loire (W) Chasselas, Blanc Fumé (Sauvignon Blanc)

Quart de Chaume (W) Chenin Blanc (Pineau de la Loire)

Quincy (W) Sauvignon Blanc

Reuilly (W) Sauvignon Blanc   (R, P) Pinot Noir, Pinot Gris

Rosé de Loire (P) Cabernet Franc, Cabernet Sauvignon, Pineau d'Aunis, Pinot Noir, Gamay, Grolleau

St-Pourçain (R) Gamay, Pinot Noir, *Gamay Teinturiers* (W) Tressallier, St-Pierre-Doré, Aligoté, Chardonnay, Sauvignon Blanc

Sancerre (W) Sauvignon Blanc   (R, P) Pinot Noir

Saumur Mousseux (W, S) Chenin Blanc, Chardonnay, Sauvignon Blanc, Cabernet Franc, Cabernet Sauvignon, Côt, Gamay, Grolleau, Pineau d'Aunis, Pinot Noir   (P, S) Cabernet Franc, Cabernet Sauvignon, Côt, Gamay, Grolleau, Pineau d'Aunis, Pinot Noir

Savennières (W) Chenin Blanc (Pineau de la Loire)

Touraine, Touraine Azay-le-Rideau, Touraine Amboise, Touraine Mesland, Touraine Pétillant (W) Chenin Blanc (Pineau de la Loire), Arbois (Menu Pineau), Sauvignon Blanc, *Chardonnay*   (R) Cabernet Franc (Breton), Cabernet Sauvignon, Côt, Pinot Noir, Pinot Meunier, Pinot Gris, Gamay Noir à Jus Blanc, Pineau d'Aunis (P) Cabernet Franc (Breton), Cabernet Sauvignon, Côt, Pinot Noir, Pinot Meunier, Pinot Gris, Gamay Noir à Jus Blanc, Pineau d'Aunis, Grolleau, *Gamay de Chaudenay, Gamay de Bouze, Gamay à Jus Coloré.*

Touraine Mousseux (W, S) Chenin Blanc (Pineau de la Loire), Arbois (Menu Pineau), *Chardonnay, Cabernet Franc (Breton), Cabernet Sauvignon, Pinot Noir, Pinot Meunier, Pinot Gris, Pineau d'Aunis, Côt,*

The Grapes Behind the Names

Grolleau (R, S) Cabernet Franc (Breton)   (P, S) Breton, Côt, Noble, Gamay, Grolleau

Valençay (R, P) Cabernet Franc, Cabernet Sauvignon, Côt, Gamay, Pinot Noir, *Gascon, Pineau d'Aunis, Gamay de Chaudenay, Grolleau*   (W) Arbois, Chardonnay, Sauvignon Blanc, *Chenin Blanc, Romorantin*

Vouvray (W) Gros Pinot (Pineau de la Loire, Chenin Blanc), Petit Pinot (Menu Pinot)

Vins de l'Orléanais (R, P) Pinot Noir, Pinot Meunier, Cabernet (W) Auvernat Blanc (Chardonnay), Auvernat Gris (Pinot Meunier)

Vins du Thouarsais (W) Chenin Blanc, *Chardonnay*   (R, P) Cabernet Franc, Cabernet Sauvignon, Gamay
Note: Gamay means Gamay Noir à Jus Blanc unless otherwise stated.

Loire Fringes

Côte Roannaise (R, P) Gamay

Côtes du Forez (R, P) Gamay

Jura

Arbois (R) Poulsard Noir, Trousseau, Pinot Noir   (W) Savagnin Blanc, Chardonnay, Pinot Blanc
Note: Pinot Noir and Pinot Blanc are not permitted for the production of *Vin de Paille*; other grapes are as above.
Arbois Rosé is made from a blend of black and white grapes.

Château-Chalon (W) Savagnin Blanc

Côte de Jura *vins jaunes* (W) Savagnin Blanc

Côtes du Jura (R) Poulsard Noir, Trousseau, Pinot Noir (W) Savagnin Blanc, Chardonnay, Pinot Blanc

Côtes du Jura Mousseux (S) Grapes as above
Côtes du Jura Rosé is made from a blend of black and white grapes.

L'Étoile (W) Chardonnay, Poulsard, Savagnin Blanc

L'Étoile Mousseux (S) As above

L'Étoile Vin Jaune (W) Savagnin Blanc
Note: Pinot Noir and Pinot Blanc are not permitted for the production of *Vin de Paille*; other grapes are as above.

Provence and Corsica

Ajaccio (P) Barbarossa, Nielluccio, Sciacarello, Vermentino Blanc, *Carignan, Cinsaut, Grenache*   (W) Ugni Blanc, Vermentino Blanc

Bandol (R) Mourvèdre, Grenache, Cinsaut, *Syrah, Carignan, Tibouren, Calitor (Pécoui Touar)*   (P) Mourvèdre, Grenache, Cinsaut, *Syrah, Carignan, Tibouren, Calitor (Pécoui Touar), Bourboulenc, Clairette, Ugni Blanc, Sauvignon Blanc*   (W) Bourboulenc, Clairette, Ugni Blanc, *Sauvignon Blanc*

Bellet, Vin de Bellet (R) Braquet, Folle Noir (Fuella), Cinsaut, *Grenache, Rolle, Roussanne, Spagnol (Mayorquin), Clairette, Bourboulenc, Chardonnay, Pignerol, Muscat Blanc à Petits Grains*   (P) Braquet, Folle Noir (Fuella), Cinsaut, *Grenache, Roussanne, Rolle, Spagnol (Mayorquin),*

Clairette, Bourboulenc, Pignerol (W) Rolle, Roussanne, Spagnol (Mayorquin), *Clairette, Bourboulenc, Chardonnay, Pignerol, Muscat Blanc à Petit Grains*

Cassis (R, P) Grenache, Carignan, Mourvèdre, Cinsaut, Barberoux, *Terret, Aramon* (W) Ugni Blanc, Sauvignon Blanc, Doucillon (Bourboulenc), Clairette, Marsanne, Pascal Blanc

Coteaux d'Aix-en-Provence, Les Baux-de-Provence (R, P) Cabernet Sauvignon, Carignan, Cinsaut, Counoise, Grenache, Mourvèdre, Syrah (W) Bourboulenc, Clairette, Grenache Blanc, Sauvignon Blanc, Sémillon, Ugni Blanc, Vermentino Blanc

Coteaux de Pierrevert, VDQS (R, P) Carignan, Cinsaut, Grenache, Mourvèdre, Oeillade (Petit), Syrah, Terret Noir (W) Clairette, Marsanne, Picpoul, Roussanne, Ugni Blanc

Coteaux Varois (R) Grenache, Syrah, Mourvèdre, *Carignan, Cinsaut, Cabernet Sauvignon* (P) Grenache, Cinsaut, *Syrah, Mourvèdre, Carignan, Tibouren* (W) Clairette, Grenache Blanc, Rolle, Sémillon, Ugni Blanc

Coteaux Varois, VDQS (R) Grenache, Syrah, Mourvèdre, *Carignan, Cinsaut, Cabernet Sauvignon* (P) Grenache, Cinsaut, *Syrah, Mourvèdre, Carignan, Tibouren, Bourboulenc Blanc, Clairette Blanc, Grenache Blanc, Ugni Blanc* (W) Clairette Blanc, Grenache Blanc, Rolle, Sémillon, Ugni Blanc

Muscat du Cap Corse (W) Muscat Blanc à Petits Grains

Palette (R, P) Mourvèdre, Grenache, Cinsaut (Plant d'Arles), *Téoulier (Manosquin), Durif, Muscat Noir de Provence, Muscat de Marseille/d'Aubagne, Muscat de Hamburg, Carignan, Syrah, Castets, Brun Fourca, Terret Gris, Petit-Brun, Tibouren, Cabernet Sauvignon* (W) Clairette à Gros Grains/Clairette à Petits Grains/Clairette de Trans/Picardan/Clairette Rosé, *Ugni Blanc, Ugni Rosé, Grenache Blanc, Muscat de Frontignan, Pascal, Terret-Bourret, Piquepoul, Aragnan, Colombard, Tokay*

Patrimonio (R, P) Nielluccio, *Grenache, Sciacarello, Vermentino Blanc* (W) Vermentino Blanc, *Ugni Blanc*

Côtes de Provence (R, P) Carignan, Cinsaut, Grenache, Mourvèdre, Tibouren, *Barberoux, Cabernet Sauvignon, Calitor (Pécoui Touar), Clairette, Roussanne du Var, Sémillon, Ugni Blanc, Vermentino Blanc (Rolle)* (W) Clairette, Sémillon, Ugni Blanc, Vermentino Blanc/Rolle

Vin de Corse (R, P) Nielluccio, Sciacarello, Grenache, *Cinsaut, Mourvèdre, Barbarossa, Syrah, Carignan, Vermentino (Malvoisie de Corse)* (W) Vermentino (Malvoisie de Corse), *Ugni Blanc (Rossola)*

RHÔNE

Châteauneuf-du-Pape (R) Grenache Noir, Cinsaut, Syrah, Mourvèdre, *Picpoul, Terret Noir, Counoise, Muscardin, Picardan, Vaccarèse, Clairette, Roussanne, Bourboulenc* (W) Grenache Blanc, Bourboulenc, Roussanne, *Clairette, Picpoul*

Châtillon-en-Diois (R, P) Gamay, *Syrah, Pinot Noir* (W) Aligoté, Chardonnay

The Grapes Behind the Names

Clairette de Die (S) Muscat Blanc à Petits Grains, *Clairette Blanche*

Condrieu, Château Grillet (W) Viognier

Cornas (R) Syrah

Coteaux de Die (W) Clairette Blanche

Coteaux de Pierrevert, VDQS (R, P) Carignan, Cinsaut, Grenache Noir, Mourvèdre (Petit) Syrah, *Oeillade, Terret Noir* (W) Clairette Blanche, Marsanne, Picpoul Blanc, Roussanne, Ugni Blanc

Côte Rôtie (R) Syrah, *Viognier*

Coteaux du Tricastin (R, P) Grenache Noir, Cinsaut, Mourvèdre Syrah, Picpoul, *Carignan, Grenache Blanc, Picpoul Blanc, Clairette Blanche, Bourboulenc, Ugni Blanc, Marsanne, Roussanne, Viognier* (W) Grenache Blanc, Picpoul Blanc, Clairette Blanche, Bourboulenc, Ugni Blanc, Marsanne, Roussanne, Viognier

Côtes du Lubéron (R, P) Grenache Noir, Syrah, Mourvèdre, Cinsaut, Carignan, *Counoise Noir, Picpoul Noir, Gamay, Pinot Noir* (W) Grenache Blanc, Clairette Blanche, Bourboulenc, Ugni Blanc, Rolle, *Roussanne, Marsanne*

Côtes-du-Rhône, Côtes-du-Rhône-Villages (R) Grenache Noir, Cinsaut, Syrah, Mourvèdre, *Terret Noir, Carignan, Counoise, Muscardin, Vaccarèse, Pinot Noir, Calitor, Gamay, Camarèse* (P) Grenache Noir, Camarèse, Cinsaut, Carignan comprising 90% of the blend, the remaining 10% could be any of these other varieties: (W) Clairette, Roussanne/Roussette, Bourboulenc, *Viognier, Picpoul, Marsanne, Grenache Blanc, Picardan, Mauzac, Pascal Blanc*

Côtes du Ventoux (R, P) Grenache Noir, Syrah, Cinsaut, Mourvèdre, Carignan, *Picpoul Noir, Counoise, Clairette, Bourboulenc, Grenache Blanc, Roussanne* (W) Clairette, Bourboulenc, Grenache Blanc, *Roussanne*

Côtes du Vivarais (R) Grenache Noir, Syrah, *Cinsaut, Carignan* (P) Grenache Noir, Cinsaut, *Syrah* (W) Clairette Blanche, Grenache Blanc, Marsanne

Crémant de Die (S) Clairette Blanche

Gigondas (R) Grenache Noir, *Syrah, Mourvèdre* (P) Grenache Noir 80%, the remainder can consist of any of the varieties under Côtes-du-Rhône

Haut-Comtat, VDQS (R, P) Grenache, *Carignan, Cinsaut, Mourvèdre, Syrah*

Hermitage, Crozes-Hermitage, St Joseph (R) Syrah, *Marsanne, Roussanne* (W) Marsanne, Roussanne

Lirac (R, P) Grenache Noir, Cinsaut, Mourvèdre, Syrah, *Carignan* (W) Clairette Blanc, Grenache Blanc, Bourboulenc, *Ugni Blanc, Picpoul, Marsanne, Roussanne, Viognier*

St-Péray (W) Roussanne (Roussette), Marsanne (S) Roussanne (Roussette), Marsanne

Tavel (P) Grenache Noir, Cinsaut, Clairette Blanche, Clairette, Picpoul, Calitor, Bourboulenc, Mourvèdre, Syrah, Carignan

Vacqueyras (R) Grenache Noir, Syrah, Mourvèdre (P) Grenache Noir, *Mourvèdre, Cinsaut*. With the exclusion of Carignan, all the

other grapes that go into Côtes-du-Rhône are also allowed
(W) Grenache Blanc, Clairette Blanc, Bourboulenc, *Marsanne Blanc,
Roussanne Blanc, Viognier*

RHÔNE FRINGES

Coteaux du Lyonnais (R) Gamay   (P) Gamay   (W) Chardonnay,
Aligoté

SAVOIE AND BUGEY

Crépy (W) Chasselas

**Vin de Bugey, Vin de Bugey Mousseux, Vin de Bugey Pétillant,
Vin de Bugey followed by a commune name** (R, P) Gamay, Pinot
Noir, Poulsard, Mondeuse plus up to 20% of the white grapes listed
here   (W) Chardonnay, Roussette, Aligoté, Mondeuse Blanche,
Jacquère, Pinot Gris, Molette

Roussette de Bugey (W) Roussette, Chardonnay

Vins de Bugey-Cerdon (W, R) Gamay, Poulsard, Pinot Noir, Pinot
Gris
Note: Varietal labelling is used for all the Bugey appellations where a wine is
exclusively from one variety.

For the commune of:
Chignin-Bergeron (W) Roussanne

For the communes of:
Marignan and Ripaille (W) Chasselas

For the commune of:
Ste-Marie d'Alloix (P, R) Gamay, Persan, Étraire de la Dui, Servanin,
Joubertin   (W) Verdesse, Jacquère, Aligoté, Chardonnay
Note: Red wines may be made with the addition of up to 20% white grapes.

Roussette de Savoie (W) Roussette, *Chardonnay*
Note: Where the designation Roussette de Savoie is followed by the name of
a commune, the wine must be 100% from Roussette.

Seyssel (W) Roussette

Seyssel Mousseux (S) Chasselas, *Roussette*

Vin de Savoie (P, R) Gamay, Mondeuse, Pinot Noir, *Persan, Cabernet
Sauvignon, Cabernet Franc, Étraire de la Dui, Servanin, Joubertin*
(W) Aligoté, Roussette, Jacquère, Chardonnay, Malvoisie, Mondeuse
Blanche, *Chasselas, Gringet, Roussette d'Ayze, Marsanne, Verdesse*
Note: An addition of up to 20% of the white grapes listed above is permitted
in the production of red Vin de Savoie.

Vin de Savoie Pétillant (W) the above white grapes plus Molette

Vin de Savoie Mousseux (S) Molette
Where the designation Vin de Savoie is followed by the name of a commune
the following grape varieties may be used:
For the communes of: Abymes, Apremont, Arbin, Ayze, Charpignat,
Chautagne, Chignin, Cruet, Montmélian, St-Jean-de-la-Porte, St-Jeoire-
Prieuré: R, W, & P as Vin de Savoie

Vin de Savoie Ayze (Pétillant & S) (W) Gringet, Roussette,
Mondeuse Blanche, *Roussette d'Ayze*

South West France

Béarn (R, P) Tannat, *Cabernet Franc (Bouchy), Cabernet Sauvignon, Fer (Pinenc), Manseng Noir, Courbu Noir* (W) Petit Manseng, Gros Manseng, *Courbu, Lauzet, Camaralet, Raffiat, Sauvignon Blanc*

Bergerac, Bergerac sec (W) Sémillon, Sauvignon Blanc, Muscadelle, Ondenc, Chenin Blanc, *Ugni Blanc*

Bergerac, Côtes de Bergerac (R, P) Cabernet Sauvignon, Cabernet Franc, Merlot, *Malbec (Côt), Fer Servadou, Merille (Périgord)*

Blanquette de Limoux Mousseux, Crémant de Limoux (W, S) Mauzac, Chardonnay, Chenin Blanc

Blanquette Méthode Ancestrale Mousseux (W, S) Mauzac

Buzet (R, P) Merlot, Cabernet Sauvignon, Cabernet Franc, Malbec (Côt) (W) Sémillon, Sauvignon Blanc, Muscadelle

Cahors (R) Malbec (Côt), Merlot, Tannat, Jurançon Noir

Côtes du Brulhois (R, P) Cabernet Sauvignon, Cabernet Franc, Merlot, Fer, Côt, Tannat

Côtes de Duras (W) Sauvignon Blanc, Sémillon, Muscadelle, Mauzac, *Rouchelin, Pineau de la Loire (Chenin Blanc) Ondenc, Ugni Blanc* (R, P) Cabernet Sauvignon, Cabernet Franc, Merlot, Malbec (Côt)

Côtes du Frontonnais/Fronton/Villaudric (R, P) Négrette, *Côt, Mérille, Fer, Syrah, Cabernet Franc, Cabernet Sauvignon, Gamay, Cinsaut, Mauzac*

Côtes du Marmandais (R, P) Cabernet Franc, Cabernet Sauvignon, Merlot, *Abouriou, Merlot (Côt), Fer, Gamay, Syrah* (W) Sauvignon Blanc, *Muscadelle, Ugni Blanc, Sémillon*

Côtes de Montravel, Haut-Montravel (W) Sémillon, Sauvignon Blanc, Muscadelle

Côtes de St-Mont, VDQS (R, P) Tannat, *Cabernet Sauvignon, Cabernet Franc, Merlot, Fer* (W) Arrufiac, Clairette, Courbu, *Gros Manseng, Petit Manseng*

Gaillac (R, P) Duras, Fer Sevadou, Gamay, Syrah, *Cabernet Sauvignon, Cabernet Franc, Merlot* (W) Len de L'el, Mauzac Rosé, Muscadelle, Ondenc, Sauvignon Blanc, Sémillon

Gaillac Premières Côtes (W) Len de L'el, Mauzac, Mauzac Rosé, Muscadelle, Ondenc, Sauvignon Blanc, Sémillon

Gaillac mousseux (doux) (W) Len de L'el, Mauzac, Mauzac Rosé, Muscadelle, Ondenc, Sauvignon Blanc, Sémillon (P) Duras, Fer Servadou, Gamay, Syrah, *Cabernet Sauvignon, Cabernet Franc, Merlot*

Irouléguy (R, P) Cabernet Sauvignon, Cabernet Franc, Tannat (W) Courbu, Manseng

Jurançon, Jurançon sec (W) Petit Manseng, Gros Manseng, *Courbu, Camaralet, Lauzet*

Limoux (W) Chardonnay

Madiran (R) Tannat, *Cabernet Sauvignon, Cabernet Franc/Bouchy, Fer (Pinenc)*

Marcillac (R, P) Fer Servadou, *Cabernet Sauvignon, Cabernet Franc, Merlot*

Monbazillac, Rosette (W) Sémillon, Sauvignon Blanc, Muscadelle

Montravel (W) Sémillon, Sauvignon Blanc, Muscadelle, Ondenc, Chenin Blanc, *Ugni Blanc*

Pacherenc du Vic Bilh (W) Arrufiac, Courbu, Gros Manseng, Petit Manseng, *Sauvignon Blanc, Sémillon*

Pécharmant (R) Cabernet Sauvignon, Cabernet Franc, Merlot, Malbec (Côt)

Saussignac (W) Semillon, Sauvignon Blanc, Muscadelle, Chenin Blanc

Tursan, VDQS (R, P) Tannat, Cabernet Sauvignon, Cabernet Franc (Bouchy), Fer Servadou (Pinenc) (W) Baroque, *Sauvignon Blanc, Petit Manseng, Gros Manseng, Claverie, Cruchinet, Raffiat, Claret de Gers, Clairette*

Vins d'Entraygues et du Fel, VDQS (R, P) Cabernet Franc, Cabernet Sauvignon, Fer Gamay, Noir à Jus Blanc, Jurançon Noir, Mouyssaguès, Négrette, Pinot Noir (W) Chenin Blanc, Mauzac

Vins d'Estaing, VDQS (R, P) Fer Servadou, Gamay Noir à Jus Blanc, Jurançon Noir, Abouriou, Merlot, Cabernet Franc, Cabernet Sauvignon, Mouyssaguès, Négrette, Pinot Noir, Duras, Castets (W) Chenin Blanc, Roussellou (St-Pierre Doré), Mauzac

Vins de Lavilledieu, VDQS (R) Négrette, Mauzac, Bordelais, Morterille (Cinsaut), Chalosse (Béquignol), *Syrah, Gamay, Jurançon Noir, Picpoul, Milgranet, Fer* (W) Mauzac, Sauvignon Blanc, Sémillon, Muscadelle, Blanquette, Ondenc, Chalosse Blanche (Claverie)

GERMANY

Liebfraumilch (W) Müller-Thurgau, Silvaner, Kerner, and/or (most unusually) Riesling must constitute 51%

GREECE

Amindeo (R) Xynomavro

Anhialos (W) Roditis, *Savatiano*

Archanes (R) Kotsifali, Mandelaria

Dafnes (R) Liatiko

Goumenissa (R) Xynomavro, *Negoska*

Limnos (W) Muscat of Alexandria

Mantinia (W) Moscophilero

Côtes de Meliton (R) Limnio, Cabernet Sauvignon, Cabernet Franc (W) Athiri, Roditis, Assyrtico, *Sauvignon Blanc, Ugni Blanc*

Naoussa (R) Xynomavro

Nemea (R) Aghiorghitiko

Paros (R) Monemvassia (Malvaria), *Mandelaria*

Patras (W) Roditis

Peza (R) Kotsifali, Mandelaria (W) Vilana

THE GRAPES BEHIND THE NAMES

Rapsani (R) Xynomavro, Krassato, Stavroto

Rhodes (R) Mandelaria   (W) Athiri

Samos (W) Muscat Blanc à Petits Grains

Santorini, Santorini Vissanto (W) Assyrtiko, *Athiri, Aidini*

Sitia (R) Liatiko

Zitsa (W) Debina

HUNGARY

Bulls Blood of Eger (Egri Bikaver) (R) Kékfrankos, Cabernet
Sauvignon, Merlot, *Kékoporto*

Tokaji (W) Furmint, Hárslevelű, *Orémus*

Tokaji Aszú Muscat (W) Furmint, Hárslevelű, Muscat Blanc à Petits
Grains

ITALY

ABRUZZI

All DOC wines are varietal

APULIA

Alezio (R, P) Negroamaro, *Malvasia Nera, Sangiovese, Montepulciano*

Brindisi (R, P) Negroamaro, *Susumaniello, Malvasia Nera, Sangiovese,
Montepulciano*

Cacc'e Mmitte di Lucera (R) Montepulciano, Uva di Troia,
Sangiovese, Malvasia Nera, Trebbiano Toscano, Bombino Bianco,
Malvasia del Chianti

Castel del Monte (W) Pampanuto (Pampanino), Chardonnay
(R) Uva di Troia, Aglianico   (P) Bombino Nero, Aglianico

Copertino (R, P) Negroamaro, *Malvasia Nera di Brindisi, Malvasia
Nera di Lecce, Sangiovese, Montepulciano*

Gioia del Colle (R, P) Primitivo, Montepulciano, Sangiovese,
Negroamaro, Malvasia Nera   (W) Trebbiano Toscano

Gravina (W) Malvasia del Chianti, Greco di Tufo, Bianco d'Alessano,
Bombino Bianco, Trebbiano Toscano, Verdeca

Leverano (P, R) Negroamaro, *Malvasia Nera di Lecce, Sangiovese,
Montepulciano, Malvasia Nera*   (W) Malvasia Bianca, *Bombino
Bianco, Trebbiano Toscano*

Lizzano (P, R) Negroamaro, *Bombino Nero, Pinot Nero, Malvasia
Nera di Lecce, Sangiovese, Montepulciano, Malvasia Nera*
(W) Trebbiano Toscano, Chardonnay, Pinot Bianco, *Malvasia, Lunga
Bianca (Malvasia), Sauvignon Blanc, Bianco d'Alessano*

Locorotondo (W) Verdeca, Bianco d'Alessano, *Fiano, Bombino,
Malvasia Toscano*

Martina (W) Verdeca, Bianco d'Alessano, *Fiano, Bombino, Malvasia
Toscano*

Matino (P, R) Negroamaro, *Malvasia Nera, Sangiovese*

Nardò (P, R) Negroamaro, *Malvasia Nera di Lecce, Montepulciano*

Orta Nova (P, R) Sangiovese, *Uva di Troia, Montepulciano, Lambrusco Maestri, Trebbiano Toscano*

Ostuni (W) Francavilla, *Bianco d'Alessano, Verdeca*

Rosso Barletta (R) Montepulciano, Sangiovese, *Malbec*

Rosso Canosa (R) Uva di Troia, *Montepulciano, Sangiovese*

Rosso di Cerognola (R) Uva di Troia, Negroamaro, *Sangiovese, Barbera, Montepulciano, Malbec, Trebbiano*

Salice Salentino (P, R) Negroamaro, *Malvasia Nera di Lecce, Malvasia Nera di Brindisi*

San Severo (W) Bombino Bianco, Trebbiano Toscano, *Malvasia Bianca*   (P, R) Montepulciano, Sangiovese

Squinzano (P, R) Malvasia Nera di Brindisi, Malvasia Nera di Lecce, Sangiovese

BASILICATA

Varietal DOCs only

CALABRIA

Cirò (P, R) Gaglioppo, *Trebbiano Toscano*

Donnici (P, R) Gaglioppo, Mantonico Nero, Greco Nero, Malvasia Bianco, Mantonico Bianco, *Pecorello*

Lamezia (W) Greco, Trebbiano, Malvasia   (P, R) Nerello, Mascalese, Nerello Cappuccio, Gaglioppo, Magliocco, Greco Nero, Marsigliana (Greco Nero)

Melissa (W) Greco Bianco, Trebbiano Toscano, Malvasia Bianca   (R) Gaglioppo, Greco Nero, Greco Bianco, Trebbiano Toscano, Malvasia

Pollino (R) Gaglioppo, Greco Nero, Malvasia Bianca, Montonico Bianco, Guarnaccia Bianca

Sant'Anna di Isola Capo Rizzuto (R, P) Gaglioppo, Nerello Mascalese, Nerello Cappuccio, Malvasia Nera, *Nocera*

San Vito di Luzzi (W) Malvasia Bianca, Greco, Trebbiano Toscano   (R, P) Gaglioppo, *Malvasia, Greco Nero, Sangiovese*

Savuto (R, P) Gaglioppo, Greco Nero, Nerello Cappuccio, Magliocco Canino, Sangiovese, Malvasia Bianca, Pecorino

Scavigna (W) Trebbiano Toscano, Chardonnay, Greco Bianco, Malvasia   (P, R) Gaglioppo, Nerello Cappuccio, Aglianico

CAMPANIA

Campi Flegrei (W) Falanghina, Biancolella, Coda di Volpe (R) Piedirosso, Aglianico, *Sciascinoso*

Capri (W) Falanghina, Greco, *Biancolella*   (R) Piedirosso

Castel san Lorenzo (R) Barbera, Sangiovese   (W) Trebbiano, Malvasia

The Grapes Behind the Names

Cilento (R) Aglianico, Piedirosso, Barbera   (P) Sangiovese, Aglianico, Primitivo, Piedirosso   (W) Fiano, Trebbiano Toscano, Greco Bianco, Malvasia

Falerno del Massico (W) Falanghina   (R) Aglianico, Piedirosso, *Primitivo, Barbera*

Guardia Sanframondi/Guardiolo (W) Malvasia Candia Bianco, Falanghina   (R) Sangiovese

Ischia (W) Forastera Bianco, Biancolella   (R) Guarnaccia, Piedirosso

Penisola Sorrentina (W) Falanghina, Biancolella, Greco (R) Piedirosso, Sciascinoso, Aglianico

Solopaca (W) Trebbiano Toscano, Falanghina, Malvasia, Coda di Volpe   (R) Sangiovese, Aglianico, *Piedirosso, Sciascinoso*

Taburno Bianco (W) Trebbiano Toscano, Falanghina (R) Sangiovese, Aglianico (S) Coda di Volpe, Falanghina

Taurasi (R) Aglianico

Vesuvio (W) Coda di Volpe, Verdeca, *Falanghina, Greco* (R) Piedirosso, Sciascinoso, *Aglianico*

Emilia-Romagna

Bianco di Scandiano (W) Sauvignon Blanc (Spergola)

Bosco Eliceo (W) Trebbiano Romagnolo, Sauvignon Blanc, Malvasia Bianca di Candia

Bosco Eliceo Sauvignon (W) Sauvignon Blanc, *Trebbiano Romagnolo*

Cagnina di Romagna (R) Refosco (Terrano)

Colli Bolognesi (W) Albana, Trebbiano Romagnolo

Colli di Parma (R) Barbera, Bonarda Piemontese, Croatina

Colli di Parma Malvasia (W) Malvasia Candia Aromatico, *Moscato Bianco*

Colli Piacentini Gutturnio (R) Barbera, Croatina (Bonarda)

Colli Piacentini Monterosso Val d'Arda (W) Malvasia di Candida Aromatica, Moscato Bianco, Trebbiano Romagnolo, Ortrugo, *Beverdino, Sauvignon Blanc*

Colli Piacentini Trebbianino Val Trebbia (W) Ortrugo, Malvasia di Candida Aromatica, Moscato Bianco, Trebbiano Romagnolo

Colli Piacentini Val Nure (W) Malvasia di Candida Aromatica, Trebbiano Romagnolo, Ortrugo

Lambrusco di Sorbara (R, P) Lambrusco di Sorbara, Lambrusco Salamino

Lambrusco Grasparossa (R, P) Lambrusco Grasparossa, *Fortana* (Uva d'Oro)

Friuli-Venezia Giulia

Aquileia (P) Merlot, Cabernet Franc, Cabernet Sauvignon, Refosco Nostrano, Refosco dal Peduncolo Rosso

Carso Terrano (R) Terrano, *Piccola Nera, Pinot Noir*

Collio Goriziano/Collio (W) Ribolla Gialla, Malvasia Istriana, Tocai Friulano, *Chardonnay, Pinot Bianco, Pinot Grigio, Riesling Italico, Riesling Renano, Sauvignon Blanc, Müller-Thurgau, Traminer Aromatico* (R) Merlot, Cabernet Franc, Cabernet Sauvignon, *Pinot Nero*

Grave del Friuli (W) Chardonnay, Pinot Bianco   (P, R) Cabernet Franc, Cabernet Sauvignon   (W, S) Chardonnay, Pinot Bianco, Pinot Nero

Isonzo (W) Tocai Friulano, Malvasia Istriana, Pinot Bianco, Chardonnay   (R) Merlot, Cabernet Franc, Cabernet Sauvignon, *Refosco dal Peduncolo Rosso, Pinot Nero*   (S) Pinot Bianco, *Chardonnay, Pinot Nero*

Latisana (P) Merlot, Cabernet Franc, Cabernet Sauvignon, Refosco Nostrano, Refosco dal Peduncolo Rosso   (W, S) Chardonnay, Pinot Bianco, Pinot Nero

Latium

Bianco Capena (W) Trebbiano, Malvasia, *Bellone, Bombino*

Cerveteri (W) Trebbiano, Malvasia, *Verdicchio, Tocai, Bellone, Bombino*   (R) Sangiovese, Montepulciano, Cesanese, *Canaiolo Nero, Carignano, Barbera*

Colli Albani (W) Malvasia, Trebbiano

Colli Lanuvini (W) Malvasia, Trebbiano

Cori (W) Malvasia, Trebbiano, *Bellone*   (R) Montepulciano, *Nero Buono di Cori, Cesanese*

Est! Est!! Est!!! di Montefiascone (W) Malvasia, Trebbiano

Frascati (W) Malvasia, Trebbiano, *Greco*

Genazzano (W) Malvasia, Bellone, Bombino, Trebbiano, Pinot Bianco   (R) Sangiovese, Cesanese

Marino (W) Malvasia, Trebbiano

Montecompatri Colonna (W) Malvasia, Trebbiano, *Bellone, Bonvino*

Velletri (W) Malvasia, Trebbiano   (R) Sangiovese, Montepulciano, Cesanese, *Bombino Nero*

Vignanello (W) Malvasia, Trebbiano   (R) Sangiovese, Ciliegiolo

Zagarolo (W) Malvasia, Trebbiano

Liguria

Cinqueterre (W) Bosco, Albarola, Vermentino

Colli di Luni (R) Sangiovese, Canaiolo, Pollera Nera, Ciliegiolo Nero   (W) Vermentino, Trebbiano Toscano

Riviera Ligure di Ponente Ormeasco (R) Dolcetto

Lombardy

Botticino (R) Barbera, Schiava Gentile, Marzemino, Sangiovese

Capriano del Colle (R) Sangiovese, Marzemino, Barbera, *Incrocio Terzi No. 1*   (W) Trebbiano di Soave, *Trebbiano Toscano*

THE GRAPES BEHIND THE NAMES

Cellatica (R) Schiava Gentile, Barbera, Marzemino, Incrocio Terzi No. 1

Colli Morenici Mantovani del Garda (R) Merlot, Rossanella (Molinara), *Negrara, Sangiovese*   (W) Trebbiano Giallo, Trebbiano Toscano, Trebbiano di Soave, Pinot Bianco, *Riesling Italico, Malvasia di Candida*

Franciacorta (R) Cabernet Franc, Barbera, Nebbiolo, Merlot (W) Pinot Bianco, Chardonnay   (S) Pinot Bianco, Chardonnay, Pinot Grigio, Pinot Nero

Lambrusco Mantovano (R) Lambrusco Viadanese, Lambrusco Maestri, Lambrusco Marani, Lambrusco Salamino, *Ancellotta, Fortana, Uva d'Oro*

Lugana (W) Trebbiano di Lugana

Oltrepò Pavese (non varietal) (R, P) Barbera, Croatina, Uva Rara, Ughetta (Vespolina), Pinot Nero

Riviera del Garda Bresciano/Garda Bresciano (W) Riesling Italico, Riesling Renano   (R, P) Groppello, Sangiovese, Marzemino, Barbera

San Colombano al Lambro (R) Croatina, Barbera, Uva Rara

San Martino della Battaglia (W) Tocai Friulano

Valcalepio (R) Merlot, Cabernet Sauvignon   (W) Pinot Bianco, Chardonnay, Pinot Grigio

Valtellina (R) Chiavennasca, *Pinot Nero, Merlot, Rossola, Pignola Valtellinese*

MARCHES

Colli Maceratesi (W) Maceratino, *Trebbiano Toscano, Verdicchio, Malvasia Toscana, Chardonnay*

Colli Pesaresi (R) Sangiovese   (W) Trebbiano

Falerio dei Colli Ascolani (W) Trebbiano, Passerina, Verdicchio, Malvasia Toscana

Focara Rosso (R) Sangiovese, *Pinot Nero*

Roncaglia Bianco (W) Trebbiano, Pinot Nero

Rosso Conero (R) Montepulciano, *Sangiovese*

Rosso Piceno (R) Montepulciano, Sangiovese, *Trebbiano, Passerina*

MOLISE

Biferno (R, P) Montepulciano, Trebbiano, Aglianico (W) Trebbiano Toscano, Bombino Bianco, Malvasia

Pentro di Isernia (W) Trebbiano Bianco, Bombino Bianco

PIEDMONTE

Asti (W, S) Moscato

Barbaresco (R) Nebbiolo Michet, Nebbiolo Lampia, Nebbiolo Rosé

Barbera d'Asti, Barbera del Monferrato (R) Barbera, *Freisa, Grignolino, Dolcetto*

Barolo (R) Nebbiolo Michet, Nebbiolo Lampia, Nebbiolo Rosé

Boca (R) Nebbiolo (Spanna), Vespolina, Bonarda Novarese

Brachetto d'Acqui (R) Brachetto, *Aleatico, Moscato Nero*

Bramaterra (R) Nebbiolo (Spanna), Croatina, Vespolina

Carema (R) Nebbiolo

Casalese (W) Cortese

Colli Tortonesi (R) Barbera, *Freisa, Bonarda Piemontese, Dolcetto* (W) Cortese

Fara (R) Nebbiolo (Spanna), *Vespolina, Bonarda*

Gabiano (R) Barbera, Freisa, Grignolino

Gattinara (R) Nebbiolo (Spanna), Vespolina, Bonarda di Gattinara

Gavi (W) Cortese

Ghemme (R) Nebbiolo (Spanna), *Vespolina, Bonarda*

Grignolino d'Asti (R) Grignolino, *Freisa*

Grignolino del Monferrato Casalese (R) Grignolino, *Freisa*

Lessona (R) Nebbiolo (Spanna), *Vespolina, Bonarda*

Loazzolo (W) Moscato

Monferrato Ciaret (R) Barbera, Bonarda, Cabernet Franc, Cabernet Sauvignon, Dolcetto, Freisa, Grignolino, Pinot Nero, Nebbiolo

Piemonte Spumante (S) Chardonnay, Pinot Bianco, Pinot Grigio, Pinot Nero

Roero (R) Nebbiolo, *Arneis* (W) Arneis

Rubino di Cantavenna (R) Barbera, *Grignolino, Freisa*

Ruchè di Castagnole Monferrato (R) Ruchè, *Barbera, Brachetto*

Sizzano (R) Nebbiolo, *Vespolina, Bonarda*

SARDINIA

Campidano di Terralba (R) Bovale Sardo, Bovale di Spagna, *Pascal di Cagliari, Greco Nero, Monica*

Mandrolisai (P, R) Bovale Sardo, Cannonau, Monica

SICILY

Alcamo (W) Catarratto Bianco Commune, Catarratto Bianco Lucido, *Damaschino, Grecanico, Trebbiano Toscano*

Cerasuolo di Vittoria (R) Frappato, Calabrese, *Grosso Nero, Nerello Mascalese*

Contessa Entellina (W) Ansonica, *Catarratto Bianco Lucido, Grecanico, Chardonnay, Sauvignon Blanc, Müller-Thurgau*

Eloro (P, R) Nero d'Avola, Frappato, Pignatello

Etna (W) Carricante, Catarratto Bianco Commune, *Trebbiano, Minnella Bianca* (P, R) Nerello Mascalese, Nerello Mantellato (Nerello Cappuccio)

Faro (R) Nerello Mascalese, Nerello Cappuccio, *Calabrese, Gaglioppo, Sangiovese, Nocera*

THE GRAPES BEHIND THE NAMES

Marsala (W) Grillo, Catarratto, Pignatello, Calabrese, Nerello Mascalese, Inzolia, Nero d'Avola, *Damaschino*

TRENTINO-ALTO ADIGE

Caldaro/Lago di Caldaro (R) Schiava, Pinot Nero, Lagrein

Casteller (R) Schiava, Merlot, Lambrusco

Klausner Leitacher (R) Schiava, *Portoghese, Lagrein*

Sorni (R) Schiava, Teroldego, Lagrein   (W) Nosiola, Müller-Thurgau, Sylvaner, Pinot Bianco

Trentino (W) Chardonnay, Pinot Bianco   (R) Cabernet Franc, Cabernet Sauvignon, Merlot (Vin Santo), Nosiola

Trento (W or P, S) Chardonnay, Pinot Bianco, Pinot Nero, Pinot Meunier

Valdadige (W) Pinot Bianco, Pinot Grigio, Riesling Italico, Müller-Thurgau, Chardonnay, Bianchetta Trevigiana, Trebbiano Toscano, Nosiola Veraccia, Garganega   (R) Schiava, Lambrusco, Merlot, Pinot Nero, Lagrein, Teroldego, Negrara

TUSCANY

Bianco della Valdinievole (W) Trebbiano Toscano, Malvasia del Chianti, Canaiolo Bianco, Vermentino

Bianco dell'Empolese (W) Trebbiano Toscano

Bianco di Pitigliano (W) Trebbiano Toscano, Greco, Malvasia Toscana, Verdello, Grechetto, *Chardonnay, Sauvignon Blanc, Pinot Bianco, Riesling Italico*

Bianco Pisano di San Torpe (W) Trebbiano

Bianco Vergine Valdichiana (W) Trebbiano

Bolgheri (W) Trebbiano, Vermentino, Sauvignon Blanc (R, P) Cabernet Sauvignon, Merlot, Sangiovese

Candia dei Colli Apuani (W) Vermentino, Albarola

Carmignano (R) Sangiovese, Canaiolo Nero, Cabernet Franc, Cabernet Sauvignon, Trebbiano Toscano, Canaiolo Bianco, Malvasia del Chianti

Carmignano/Barco Reale di Carmignano (R) Sangiovese, Canaiolo Nero, *Cabernet Franc, Cabernet Sauvignon, Trebbiano Toscano, Canaiolo Bianco, Malvasia*

Chianti (R) Sangiovese, Canaiolo Nero, Trebbiano, Malvasia del Chianti

Colli dell'Etruria Centrale (R) Sangiovese, *Canaiolo Nero, Trebbiano, Malvasia, Cabernet Franc, Cabernet Sauvignon, Merlot* (W) Trebbiano Toscano, *Malvasia del Chianti, Pinot Bianco, Pinot Grigio, Chardonnay, Sauvignon Blanc*

Colline Lucchese (R) Sangiovese, Canaiolo, Ciliegiolo, Colorino, Trebbiano, Vermentino   (W) Trebbiano Toscano, Greco, Grechetto, Vermentino Bianco, Malvasia

Elba (W) Trebbiano (Procanico)   (R) Sangiovese (Sangioveto)

Montecarlo (B) Trebbiano, Semillon, Pinot Grigio, Pinot Bianco, Vermentino, Sauvignon Blanc, Roussanne (R) Sangiovese, Canaiolo Nero, Ciliegiolo, Colorino, Malvasia Nera, Syrah, Cabernet Franc, Cabernet Sauvignon, Merlot

Monteregio di Massa Marittima (R) Sangiovese (W) Trebbiano, Vermentino, Malvasia, Ansonica

Montescudaio (W) Trebbiano, Vermentino, Malvasia (R) Sangiovese, Trebbiano Toscano, Malvasia

Parrina (R, P) Sangiovese (W) Trebbiano, Ansonica, Chardonnay

Pomino (R) Sangiovese, Canaiolo, Cabernet Sauvignon, Cabernet Franc, Merlot (W) Pinot Bianco, Chardonnay, Trebbiano

Rosso di Montalcino (R) Sangiovese

Rosso di Montepulciano (R) Sangiovese, Canaiolo Nero

Sassicaia (R) Cabernet Sauvignon

Val d'Arbia (W) Trebbiano, Malvasia, Chardonnay

Val di Cornia (W) Trebbiano, Vermentino, *Malvasia, Ansonica, Biancame, Clairette, Pinot Bianco, Pinot Grigio* (R) Sangiovese, *Canaiolo, Ciliegiolo, Cabernet Sauvignon, Merlot*

Vino Nobile di Montepulciano (R) Sangiovese, Canaiolo Nero

Vino Santo Occhio di Pernice (P) Sangiovese, Merlot

UMBRIA

Colli Altotiberini (W) Trebbiano Toscano, Malvasia del Chianti (R) Sangiovese, Merlot, Trebbiano Toscano, Malvasia del Chianti Nero

Colli Amerini (W) Trebbiano Toscano, Grechetto, Verdello, Garganega, Malvasia Toscano (R) Sangiovese

Colli del Trasimeno (R) Sangiovese, Ciliegiolo, Gamay, Malvasia del Chianti, Trebbiano Toscana (W) Trebbiano Toscana, Malvasia del Chianti, Verdicchio Bianco, Verdello, Grechetto

Colli Perugini (R) Sangiovese, Montepulciano, Ciliegiolo, Barbera, Merlot (W) Trebbiano Toscano, Verdicchio, Grechetto, Garganega, Malvasia del Chianti

Montefalco (W) Grechetto, Trebbiano Toscano (R) Sangiovese, Sagrantino

Orvieto (W) Trebbiano Toscano (Procanico), Verdello, Grechetto, Canaiolo Bianco (Drupeggio), Malvasia Toscana

Torgiano (W) Trebbiano Toscano, Grechetto (R) Sangiovese, Canaiolo, *Trebbiano Toscano, Ciliegiolo, Montepulciano* (S) Chardonnay, Pinot Nero

Torgiano Riserva (R) Sangiovese, Canaiolo, *Trebbiano Toscano, Ciliegiolo, Montepulciano*

VALLE D'AOSTA

Arnad-Montjovet (R) Nebbiolo, *Dolcetto, Vien de Nus, Pinot Nero, Neyret, Freisa*

Chambave (R) Petit Rouge, *Dolcetto, Gamay, Pinot Nero* (W) Moscato

The Grapes Behind the Names

Donnaz (R) Nebbiolo (Picutener), Freisa, Neyret

Enfer d'Arvier (R) Petit Rouge, Vien de Nus, Neyret, Dolcetto, Pinot Nero, Gamay

Nus (R) Vien de Nus, Petit Rouge, Pinot Nero   (W) Malvoisie

Torrette (R) Petit Rouge, *Gamay, Pinot Nero, Fumin, Vien de Nus, Dolcetto, Mayolet, Premetta*

Valle d'Aosta (W) Müller-Thurgau, Pinot Grigio, Petite Arvine, Chardonnay, Blanc de Morgex   (R) Petit Rouge, Chambave, Dolcetto, Gamay, Pinot Nero, *Premetta*, Fumin

VENETO

Bardolino (R) Corvina, Rondinella, Molinara, Negrara, *Rossignola, Barbera, Sangiovese, Garganega*

Bianco di Custoza (W) Trebbiano Toscano, Garganega, Tocai Friulano, Cortese, Malvasia, Pinot Bianco, Chardonnay, Riesling Italico

Breganze (W) Tocai, *Pinot Bianco, Pinot Grigio, Riesling Italico, Sauvignon Blanc, Vespaiolo*   (R) Merlot, *Groppello Gentile, Cabernet Franc, Cabernet Sauvignon, Pinot Nero, Freisa*

Colli Berici Chardonnay (W) Chardonnay, *Pinot Bianco*

Colli Berici Garganega (W) Garganega, *Trebbiano di Soave (Trebbiano Nostrano)*

Colli Berici Pinot Bianco (W) Pinot Bianco, *Pinot Grigio*

Colli Berici Sauvignon (W) Sauvignon Blanc, *Garganega*

Colli Berici Spumante (S) Garganega, *Pinot Bianco, Pinot Grigio*

Colli Berici Tocai Italico (W) Tocai Italico, *Garganega*

Colli Berici Tocai Rosso (R) Tocai Rosso, *Garganega*

Colli di Conegliano (W) Incrocio Manzoni 6.0.13., Pinot Bianco, Chardonnay, *Sauvignon Blanc, Riesling Renano*   (R) Cabernet Franc, Cabernet Sauvignon, Marzemino, Merlot, *Incrocio Manzoni 2.15*

Colli Euganei (W) Garganega, Prosecco (Serprina), Tocai, Friulano, Sauvignon Blanc, *Pinella, Pinot Bianco, Riesling Italico, Chardonnay*   (R) Merlot, Cabernet Franc, Cabernet Sauvignon, Barbera, Raboso Veronese

Gambellara (W) Garganega

Lessini Durello (W) Durella, *Garganega, Trebbiano di Soave, Pinot Bianco, Pinot Nero, Chardonnay*

Fior d'Arancio (W) Moscato Giallo

Montello e Colli Asolani (R) Merlot, Cabernet Franc, Cabernet Sauvignon

Montello e Colli Asolani Cabernet (R) Cabernet Franc, Cabernet Sauvignon, *Malbec*

Montello e Colli Asolani Chardonnay (W) Chardonnay, *Pinot Bianco, Pinot Grigio*

Montello e Colli Asolani Merlot (R) Merlot, *Cabernet Sauvignon, Cabernet Franc*

Montello e Colli Asolani Pinot Bianco (W) Pinot Bianco, *Chardonnay, Pinot Grigio*

Montello e Colli Asolani Pinot Grigio (W) Pinot Grigio, *Chardonnay, Pinot Bianco*

Montello e Colli Asolani Prosecco (W) Prosecco, *Chardonnay, Pinot Bianco, Pinot Grigio, Riesling Italico, Bianchetta Trevigiana*

Prosecco di Conegliano-Valdobbiadene/Prosecco di Conegliano/Prosecco di Valdobbiadene (W) Prosecco, *Verdiso, Pinot Bianco, Pinot Grigio, Chardonnay*

Refrontolo Passito (R) Marzemino

Soave (W) Garganega, Pinot Bianco, Chardonnay, Trebbiano

Torchiato di Fregona (W) Prosecco, Verdiso, *Boschera*

Valpolicella (R) Corvina Veronese, Rondinella, Molinara

PORTUGAL

Alcobaça (IPR) (R) Periquita, Baga, Trincadeira   (W) Vital, Tamarez, Fernão Pires, Malvasia, Arinto

Alenquer (IPR) (R) Camarate, Mortágua, Periquita, Preto Martinho, Tinta Miúda   (W) Vital, Jampal, Arinto, Fernão Pires

Almeirim (IPR) (R) Castelão Nacional, Poeirinha (Baga), Periquita, Trincadeira Preta   (W) Fernão Pires, Arinto, Rabo de Ovelha, Talia, Trincadeira das Pratas, Vital

Arrábida (IPR) (R) Castelão Francês (Periquita), Alfrocheiro, Cabernet Sauvignon   (W) Fernão Pires, Arinto, Moscatel de Setúbal, Rabo de Ovelha, Roupeiro

Arruda (IPR) (R) Camarate, Trincadeira, Tinta Miúda   (W) Vital, Jampal, Fernão Pires

Bairrada (R) Baga, Castelão Francês, Tinta Pinheira   (W) Maria Gomes, Bical, Rabo de Ovelha

Biscoitos (IPR) (W) Arinto, Terrantez, Verdelho

Borba (R) Aragonez, Periquita, Trincadeira   (W) Perrum, Rabo de Ovelha, Roupeiro, Tamarez

Bucelas (W) Arinto, Esgana Cão

Carcavelos (R, W) Galego Dourado, Boal, Arinto, Trincadeira Torneiro, Negra Mole

Cartaxo (IPR) (R) Castelão Nacional, Periquita, Preto Martinho, Trincadeira Preta   (W) Fernão Pires, Talia, Trincadeira das Pratas, Vital, Arinto

Castelo Rodrigo (IPR) (R) Bastardo, Marufo, Rufete, Touriga Nacional   (W) Codo, Arinto do Dão, Arinto Gordo, Fonte-Cal

Chamusca (R) Castelão Nacional, Periquita, Trincadeira Preta   (W) Fernão Pires, Talia, Arinto, Trincadeira das Pratas, Vital

Chaves (IPR) (R) Tinta Amarela, Bastardo, Tinta Carvalha (W) Gouveio, Malvasia Fina, Códega, Boal

Colares (R) Ramisco   (W) Arinto, Jampal, Galego Dourado, Malvasia

The Grapes Behind the Names

Coruche (IPR) (R) Periquita, Preto Martinho, Trincadeira Preta (W) Fernão Pires, Talia, Trincadeira das Pratas, Vital

Cova da Beira (IPR) (R) Jaén, Marufo, Periquita, Rufete, Tinta Amarela (W) Pérola, Rabo de Ovelha, Arinta de Dão, Arinto Gordo

Dão (R) Alfrocheiro Preto, Bastardo, Jaén, Tinta Pinheira, Tinta Roriz, Touriga Nacional (W) Encruzado, Assario Branco, Barcelo, Borrado das Moscas, Cerceal, Verdelho

Douro (R) Touriga Nacional, Touriga Francesa, Tinta Roriz, Tinta Barroca, Tinto Cão, Tinta Amarela, Mourisco Tinto, Bastardo (W) Gouveio, Viosinho, Rabigato, Malvasia Fina, Donzelinho

Encostas d'Aire (IPR) (R) Periquita, Baga, Trincadeira Preta (W) Fernão Pires, Arinto, Tamarez, Vital

Encostas de Nave (IPR) (R) Touriga Nacional, Touriga Francesa, Tinta Barroca, Mourisco Tinto (W) Malvasia Fina, Folgosão, Gouveio

Évora (IPR) (R) Periquita, Trincadeira, Aragonez, Tinta Caida (W) Arinto, Rabo de Ovelha, Roupeiro, Tamarez

Graciosa (IPR) (W) Arinto, Boal, Fernão Pires, Terrantez, Verdelho

Granja-Amareleja (IPR) (R) Moreto, Periquita, Trincadeira (W) Mantuedo, Rabo de Ovelha, Roupeiro

Lafões (IPR) (R) Amaral, Jaén (W) Arinto, Cerceal

Lagoa (R) Negra Mole, Periquita (W) Crato Branco

Lagos (R) Negra Mole, Periquita (W) Boal Branco

Madeira (R) Tinta Negra Mole, Bastardo, Malvasia Roxa, Verdelho Tinto (W) Sercial, Verdelho, Boal, Malvasia, Terrantez

Moura (IPR) (R) Alfrocheiro, Moreto, Periquita, Trincadeira (W) Antão Vaz, Fernão Pires, Rabo de Ovelha, Roupeiro

Óbidos (IPR) (R) Periquita, Bastardo, Camarate, Tinta Miúda (W) Vital, Arinto, Fernão Pires, Rabo de Ovelha

Palmela (IPR) (R) Periquita, Alfrocheiro, Espadeiro (W) Fernão Pires, Arinto, Rabo de Ovelha, Moscatel de Setúbal, Tamarez

Pico (IPR) (W) Arinto, Terrantez, Verdelho

Pinhel (IPR) (R) Bastardo, Marufo, Rufete, Touriga Nacional (W) Codo, Arinto do Dào, Fonte-Cal

Planalto Mirandês (IPR) (R) Touriga Nacional, Touriga Francesa, Tinta Amarela, Mourisco Tinto, Bastardo (W) Gouveio, Malvasia Fina, Rabigato, Viosinho

Portalegre (IPR) (R) Aragonez, Grand Noir, Periquita, Trincadeira (W) Arinto, Galego, Roupeiro, Assário, Manteudo, Fernão Pires

Portimão (R) Negra Mole, Periquita (W) Crato Branco

Port (R) Touriga Francesa, Touriga Nacional, Bastardo, Mourisco, Tinto Cão, Tinta Roriz, Tinta Amarela, Tinta Barroca (W) Gouveio (Verdelho), Malvasia Fina, Rabigato (Rabo di Ovelha), Viosinho, Donzelinho, *Códega*

Redondo (R) Aragonez, Moreto, Periquita, Trincadeira (W) Fernão Pires, Rabo de Ovelha, Mantendo, Roupeiro, Tamarez

Reguengos (R) Aragonez, Moreto, Periquita, Trincadeira
(W) Manteudo, Perrum, Rabo de Ovelha, Roupeiro

Santarém (IPR) (R) Castelão Nacional, Periquita, Preto Martinho,
Trincadeira Preta   (W) Fernão Pires, Arinto, Rabo de Ovelha,
Talia, Trincadeira das Pratas, Vital

Setúbal (W) Moscatel de Setúbal (Muscat of Alexandria), Moscatel
Roxo, Tamarez, Arinto, Fernão Pires

Tavira (R) Negra Mole, Periquita   (W) Crato Branco

Tomar (IPR) (R) Castelão Nacional, Baga, Periquita   (W) Fernão
Pires, Arinto, Malvasia, Rabo de Ovelha, Talia

Torres Vedras (IPR) (R) Camarate, Mortágua, Periquita, Tinta
Miúda   (W) Vital, Jampal, Rabo de Ovelha, Arinto, Fernão Pires,
Seara Nova

Valpaços (IPR) (R) Touriga Nacional, Touriga Francesa, Tinta Roriz,
Tinta Amarela, Tinta Carvalha, Mourisco Tinto, Cornifesto,
Bastardo   (W) Códega, Fernão Pires, Gouveio, Malvasia Fina, Boal,
Rabigato

Varosa (R) Alvarelhão, Tinta Barroca, Tinta Roriz, Touriga Francesa,
Touriga Nacional   (W) Malvasia Fina, Arinto, Borrado das Moscas,
Cerceal, Gouveio, Fernão Pires, Folgosão

Vidigueira (IPR) (R) Alfrocheiro, Moreto, Periquita,
Trincadeira   (W) Antão Vaz, Mantuedo, Perrum, Rabo de Ovelha,
Roupeiro

Vinho Verde (R) Vinhão (Sousão), Espadeiro, Azal Tinto, Borraçal,
Brancelho, Pedral   (W) Loureiro, Trajadura, Padernã, Azal,
Avesso, Alvarinho

ROMANIA

Cotnari (W) Grasă, Tămîioasă, Francusa, Fetească Albă

SPAIN

Alella (W) Pansá Blanca (Xarel-lo), Garnacha Blanca, Macabeo,
Chardonnay, Pansá Rosado, Chenin Blanc   (R) Ull de Llebre
(Tempranillo), Garnacha Tinta, Garnacha Peluda

Alicante (R) Monastrell, Garnacha Tinta, Bobal   (W) Merseguera,
Moscatel Romano, Verdil

Almansa (R) Monastrell, Cencibel, Garnacha Tintorera
(W) Merseguera

Ampurdán-Costa Brava (R) Garnacha Tinta, Cariñena, Cabernet
Sauvignon, Merlot, Tempranillo, Garnacha, Syrah   (W) Macabeo,
Garnacha Blanca, Chenin Blanc, Riesling, Muscat, Gewürztraminer,
Macabeo, Chardonnay, Parellada, Xarel-lo

Bierzo (R) Mencía, Garnacha Tintorera   (W) Godello, Doña
Blanca, Malvasía, Palomino

Binissalem (R) Manto Negro, Callet, Tempranillo, Monastrell
(W) Moll, Parellada, Macabeo

Bizkaiko Txakolina/Chacoli de Vizcaya (W) Hondarrabi Zuri, Folle Blanche (Gros Plant)   (R) Hondarrabi Beltza

Bullas (R) Monastrell, Tempranillo   (W) Macabeo, Airén

Calatayud (R) Garnacha Tinta, Tempranillo, Cariñena, Juan Ibáñez, Monastrell   (W) Viura, Garnacha Blanca, Moscatel Romano, Malvasía

Campo de Borja (R) Garnacha, Tempranillo   (W) Macabeo

Cariñena (R) Garnacha, Tempranillo, Cariñena, Juan Ibáñez, Monastrell, Cabernet Sauvignon   (W) Viura, Garnacha Blanca, Parellada, Moscatel Romano

Cava (S) Xarel-lo, Parellada, Macabeo, Chardonnay

Chacoli de Guetaria (W) Hondarrabi Zuri   (R) Hondarrabi Beltza

Cigales (R) Tempranillo, Garnacha   (W) Verdejo, Viura, Palomino, Albillo
Note: All white grapes may be used for rosé wines.

Conca de Barbera (W) Macabeo, Parellada   (R) Trepat, Garnacha, Ull de Llebre (Tempranillo), Cabernet Sauvignon

Condado de Huelva (W) Zalema, Palomino, Garrido Fino, Moscatel

Costers del Segre (W) Chardonnay, *Macabeo, Parellada, Xarel-lo, Garnacha Blanca*   (R) Tempranillo, Cabernet Sauvignon, Merlot, *Monastrell, Trepat, Mazuelo (Cariñena), Garnacha Tinta*

Jumilla (R) Monastrell, Garnacha Tintorera, Cencibel   (W) Merseguera, Airén, Pedro Ximénez

Lanzarote (W) Burrablanca, Breval, Diego, Listán Blanca, Malvasía, Moscatel, Pedro Ximénez   (R) Listán Negra, Negramoll

La Mancha (R) Cencibel, *Garnacha, Moravia*   (W) Airén, *Pardillo, Verdoncho, Macabeo*

Méntrida (R) Garnacha, Tinto Madrid, Cencibel

Montilla-Moriles (W) Pedro Ximénez, Lairén (Airén), *Baladi, Torrontés, Moscatel*

Navarra (R) Tempranillo, Garnacha Tinta, Cabernet Sauvignon, Merlot, *Mazuelo, Graciano*   (W) Viura, *Moscatel de Grano Menudo (Muscat de Frontignan), Malvasia Riojana, Chardonnay, Garnacha Blanca*

La Palma (W) Albillo, Bastardo Blanco, Bermejuela, Bujariego, Burrablanca, Forastera Blanca, Bual, Listán Blanco, Malvasía, Moscatel, Pedro Ximénez, Sabro, Torrontés, Verdello   (R) Almuñeco (Listán Negro), Bastardo Negro, Malvasía Rosada, Moscatel Negro, Negramoll, Tintilla

Penedès (R) Tempranillo, Garnacha Tinta, Cabernet Franc, Merlot, Pinot Noir, Cabernet Sauvignon, Monastrell, Cariñena, Samsó   (W) Parellada, Xarel-lo, Macabeo, Subirat Parent, Gewürztraminer, Muscat d'Alsace, Chardonnay, Sauvignon, *Chenin Blanc, Riesling*

Priorato (R) Garnacha Tinta, *Garnacha Peluda, Cariñena, Cabernet Sauvignon*   (W) Garnacha Blanca, *Macabeo, Pedro Ximénez, Chenin Blanc*

Rias Baixas (W) Albariño, *Treixadura, Loureira Blanca, Caiño Blanco Torrontés*   (R) Brancellao, Caiño Tinto, Espadeiro, Loureira Tinta, Mencía, Sousón

Ribeira Sacra (W) Albariño, Loureira, Godello, Doña Blanca, Torrontés, *Palomino* (R) Mencía, Brancellao, Sousón, Merenzao

Ribeiro (W) Treixadura, Loureira, Albariño, *Jerez (Palomino), Torrontés, Godello, Macabeo, Albillo* (R) Caiño, Garnacha (Alicante), Ferrón, Sousón, Mencía, Tempranillo, Brancellao

Ribera del Duero (R) Tinto Fino / Tinta del País, *Garnacha Tinta (Tinto Aragonés), Cabernet Sauvignon, Merlot, Malbec, Albillo*

Rioja (R) Tempranillo, Garnacha, *Graciano, Mazuelo, and Cabernet Sauvignon (experimental)* (W) Viura, *Malvasia Riojana, Garnacha Blanca*

Rueda (W) Verdejo, Viura, *Sauvignon Blanc, Palomino Fino*

Somontano (R) Moristel, Tempranillo, Cabernet Sauvignon, Merlot (W) Viura, Alcañón, Chardonnay, Pinot Noir, Chenin Blanc, Gewürztraminer

Tacoronte-Acentejo (R) Listán Negro, Negramoll (W) Malvasía, Moscatel Blanco, Listán (Palomino)

Tarragona (R) Garnacha Tinta, Cariñena, Ull de Llebre (Tempranillo), *Cabernet Sauvignon, Merlot* (W) Macabeo, Xarel-lo, Parellada, Garnacha Blanca, *Chardonnay, Muscat*

Terra Alta (W) Garnacha Blanca, Macabeo, *Chardonnay, Colombard* (R) Cariñena, Garnacha Tinta, Garnacha Peluda, *Pinot Noir, Pinot Meunier, Cabernet Sauvignon, Merlot*

Toro (R) Tinto de Toro (Tempranillo), *Garnacha Tinta, Cabernet Sauvignon* (W) Malvasia, Verdejo Blanco

Utiel-Requena (R) Tempranillo, Bobal, Garnacha Tinta, *Cabernet Sauvignon* (W) Macabeo, Merseguera, *Planta Nova, Chardonnay*

Valdeorras (R) Mencía, Garnacha, Gran Negro, Maria Ordoña (Merenzao) (W) Godello, Palomino, Valenciana (Doña Blanca), Lado

Valdepeñas (R) Cencibel (W) Airén

Valencia (W) Merseguera, Malvasia Riojana, Planta Fina, Pedro Ximénez, Moscatel Romano, Macabeo, Tortosí (Bobal Blanco) (R) Monastrell, Garnacha Tintorera, Garnacha Tinta, Tempranillo, Forcayat

Vinos de Madrid (R) Tinto Fino, Garnacha (W) Malvar, Airén, Albillo

Ycoden-Daute-Isora (W) Bastardo Blanco, Bermejuela, Forastera Blanca, Bual, Listán Blanco, Malvasía, Moscatel, Pedro Ximénez, Sabró, Torrontés, Verdello, Vijariego (R) Bastardo Negro, Listán Negra, Malvasía Rosada, Moscatel Negra, Negramoll, Tintilla, Vijariego Negro

Yecla (R) Monastrell, Garnacha, *Cabernet Sauvignon, Tempranillo* (W) Merseguera, Verdil

SWITZERLAND

Dôle (R) Pinot Noir, *Gamay*

Goron (Valais), Salvagnin (Vaud) (R) Pinot Noir, Gamay

L'Œil-de-Perdrix de Neuchâtel (P) Pinot Noir

Acknowledgements and Sources

A prime reference for this pocket book has been Patrick W. Fegan of the Chicago Wine School. He is the only person I know who is as excited by grape variety statistics as I am and he has been kind enough to give me access to his painstakingly assembled database, although he should not be blamed for errors or omissions as I have reinterpreted some of the data.

Everyone writing about ampelography, the science of vine identification by description, owes a huge debt to Pierre Galet of Montpellier, who has written more than any other modern authority on this subject. Fortunately, his work is being continued by an increasing band of experts in France and all over the expanding wine world.

Other individuals who have passed on their knowledge most selflessly and energetically include Victor de la Serna on Spain, Daniel Thomases on Italy, Nico Manessis on Greece, and, as always, Richard Smart on general matters viticultural.

The London office of the Wine & Spirit Education Trust did a sterling job of sorting out the Grapes Behind the Names section, uncovering DOCs and IPRs previously unexposed to international scrutiny.

I have also referred, sometimes extensively, to the following:

ALCALDE, ALBERTO J., *Cultivares Viticolas Argentinas* (Mendoza, 1989).

AMBROSI, H., DETTWEILER, E., RÜHL, E. H. R., SCHMID, J. S., and SCHUMANN, F. S., *Farbatlas Rebsorten* (Ulmer, 1994).

ANDERSON, B., *The Wine Atlas of Italy* (London, 1990).

BOUCHARD, A., 'Notes ampelographiques rétrospectives sur les cépages de la généralité de Dijon', *Bulletin de la Société des Viticulteurs de France* (Paris, 1899).

DAUREL J., *Les Raisins de cuve de la Gironde et du Sud-Ouest de la France* (Bordeaux, 1892).

ENJALBERT, H., *Les Grands Vins de St-Émilion, Pomerol et Fronsac* (Paris, 1983).

FEGAN, P. W., *The Vineyard Handbook* (Chicago, 1992).

GALET, P., *Cépages et vignobles de France* (2nd edn.), Montpellier, 1990).

—— 'La Culture de la vigne aux États-Unis et au Canada', *France viticole* (Sept.-Oct. 1980 and Jan.-Feb. 1981).

—— *A Practical Ampelography*, trans. and adapted by L. T. Morton (Ithaca, NY, 1979).

GLEAVE, D., *The Wines of Italy* (London, 1989).

LIVINGSTONE-LEARMONTH, J., *The Wines of the Rhône* (3rd edn., London, 1992).

MANESSIS N., *The Greek Wine Guide* (2nd edn., Corfu, 1994).

MAYSON, R., *Portugal's Wines & Wine Makers* (London, 1992).

ODART, A., *Traité des Cépages* (3rd edn., Paris, 1854).

PLATTER, J., *South African Wine Guide* (Stellenbosch, annual).

RADFORD, J., *The Wines from Spain Guide to the Wines from Spain* (London, 1993).

ROBINSON, J., *The Oxford Companion to Wine* (Oxford, 1995).

—— *Vines, Grapes and Wines* (London, 1986).

SLOAN, JOHN C., *The Surprising Wines of Switzerland* (London, 1995).

SULLIVAN C. L., 'A Viticultural Mystery Solved', *California History*, 57 (Summer 1978), 115–29.

A REQUEST FOR MORE INFORMATION

If you have read this far, you obviously share my interest in the vine varieties of the wine world. You may also share my concern that no grape should be abandoned through lack of interest. One ambition in compiling this book has been to assemble a complete register of every single grape and vine variety of relevance to wine drinkers and producers — but some may well have escaped my notice.

If you come across a wine grape variety that does not appear in this book, I would be very grateful if you would take the trouble to let me know, sending as many details as possible to the address below. I can then make further investigations and, if necessary, make adjustments to future editions.

Vine growers and label copy-writers tend to be bad at spelling, however, so beware apparently novel grape names which are simply casually-spelt synonyms. And bear in mind that I have deliberately omitted grapes which are normally used for the fruit basket or for drying.

With very many thanks in advance,

Jancis R

Jancis Robinson
c/o Alysoun Owen
Reference Department
Oxford University Press
Walton Street
Oxford OX2 6DP
United Kingdom

NOTES

THE OXFORD COMPANION TO WINE

From the novice to the connoisseur, wine lovers will be enlightened, entertained, and enchanted with this complete one-volume guide to the world of wine. Edited under the keen supervision of Jancis Robinson, Master of Wine, and 'perhaps the most gifted of all wine writers writing today' according to *The Wine Advocate*, the *Companion* includes over 3,000 entries by some 70 leading authorities from around the world. Lavishly illustrated with hundreds of illustrations, photographs, and maps, it is the ultimate A-Z reference for today's wine lover.

THE OXFORD COMPANION TO WINE
INCLUDES IN-DEPTH COVERAGE OF:

The art of wine appreciation

How to taste and score wine

The history of wine

Wines by region

Who's Who in the wine world

Hundreds of grape varieties

Distilled and fortified wines

The science of wine-making

Amateur wine-making

The wine trade

Building a wine cellar

. . . and much more